Thomas Kuhn

Philosophy Now

Series Editor: John Shand

This is a fresh and vital series of new introductions to today's most read, discussed and important philosophers. Combining rigorous analysis with authoritative exposition, each book gives clear and comprehensive access to the ideas of those philosophers who have made a truly fundamental and original contribution to the subject. Together the volumes comprise a remarkable gallery of the thinkers who have been at the forefront of philosophical ideas.

Published

Thomas Kuhn
Alexander Bird

John Searle
Nick Fotion

Charles Taylor
Ruth Abbey

Thomas Kuhn

Alexander Bird

PRINCETON UNIVERSITY PRESS
PRINCETON, NEW JERSEY

Published in North and South America by
Princeton University Press, 41 William Street,
Princeton, New Jersey 08540. All rights reserved.

First published in 2000 by Acumen
Acumen Publishing Limited
15a Lewins Yard
East Street
Chesham
Bucks HP5 1HQ, UK

Library of Congress Catalog Card Number 00-110976

ISBN 0-691-05709-5 (hardcover)
ISBN 0-691-05710-9 (paperback)

Designed and typeset in Century Schoolbook
by Kate Williams, Abergavenny.
Printed and bound by Biddles Ltd., Guildford and King's Lynn.

www.pup.princeton.edu

10   9   8   7   6   5   4   3   2   1

Contents

For Lucasta Miller, Ian Bostridge and Oliver.

Preface

In the first two-thirds of the twentieth century the philosophy of science was, alongside logic and the philosophy of mathematics, central to what we now call the "analytic" tradition in philosophy. For the logical positivists in particular, the distinction between philosophy of science and the rest of philosophy scarcely existed. Science was the paradigm of *a posteriori* knowledge, knowledge gained through the senses (logic and mathematics were paradigmatic of *a priori* knowledge, knowledge gained from pure reflection). Therefore, insofar as the central questions of philosophy were concerned with epistemology (the theory of knowledge), the philosophical understanding of science was essential to understanding how knowledge of the world was possible. Furthermore, just as science was regarded as paradigmatic of empirical knowledge, scientific language was correspondingly taken to be characteristic of any language used to talk about the world. And so philosophy of science remained central to the logical positivists, even when their concerns shifted from the explicitly epistemological to the seemingly new field of the philosophical analysis of language.

The philosophical position of the philosophy of science in the last third of the twentieth century was rather different. Its concerns were clearly distinct from those of core epistemology and the philosophy of language. For example, while the main preoccupation of orthodox epistemologists was the search for the proper definition of (personal) knowledge, philosophers of science ignored this altogether, being more concerned with the nature of change in theory preference among groups of scientists. More generally, philosophers of science now took it for granted that their problems and insights would come not from other parts of philosophy but rather from the history of science. This shift was reflected institutionally in the formation at many English-speaking universities of departments of history and philosophy of science.

This new historical direction in the philosophy of science had predominantly Thomas Kuhn to thank. Others certainly played a part in forming the new-style philosophy of science – Paul Feyerabend and Imre Lakatos in particular – but even in Lakatos' case that is in large measure attributable to his reaction to Kuhn. Merely to have been responsible for this change is to mark Kuhn as one of the historically most significant philosophers of the twentieth century, and when we think of his influence beyond the philosophy of science, not only in the history of science but in a wide variety of areas in the social sciences and humanities, our appreciation of his significance must increase accordingly. Relative to this enormous influence the scope of this book is limited. It is limited in one important respect (and in two less important ones). First, it is largely limited to the philosophy of science. I do not discuss at length Kuhn's historical work nor do I consider in detail his influence outside the history and philosophy of science. There is however a sense in which this limitation is only partial. For I claim that Kuhn's *The Structure of Scientific Revolutions* (1962) is not primarily a philosophy text. Rather it is a work in what I call "theoretical history".[1] The central claim of his book is that the history of science displays a certain pattern and the main subsidiary claim is that this pattern may be explained by reference to the institutional structure of science, specifically the way professional scientists base their research on certain objects of consensus that Kuhn calls 'paradigms'. As they stand neither of these claims seems explicitly philosophical. Analogous theories may be found in the natural and social sciences.[2] Émile Durkheim claimed that he found patterns in certain social statistics, such as variations in suicide rates in different countries, religions, professions and so forth, and explained them by reference to his notion of "anomie". Just as any critical discussion of Durkheim must be partly – indeed primarily – sociological, a discussion of Kuhn's theory must have a significant historical content. Whether the pattern claimed exists and whether the explanation provided is plausible will be mainly non-philosophical questions.

As in Durkheim's case, only more so in Kuhn's, theory is nonetheless philosophically highly significant. His theory, both in outline and in detail, presents a challenge to certain theses widely held at the time, and in mounting this challenge Kuhn puts forward philosophical claims of his own. It is with this aspect of his work that this book is concerned above all. Furthermore, the concern with his philosophical ideas is itself philosophical. While I do aim to put Kuhn's ideas into their historical context, this is at least as much a heuristic to help clarify what those ideas are as an exercise in the history of ideas. This is the second limitation of the book. I am interested in why Kuhn thought what he did primarily as a guide to understanding accurately what he did think. I am correspondingly rather less interested in what Kuhn's followers, supporters and critics have said, except where that may itself illuminate interesting aspects of Kuhn's thought. My aims are first to expound what Kuhn thinks on a topic and then to examine the views thus expounded to see whether we should believe them or not. This is a critical, philosophical engagement with Kuhn's thought and the main reason why I have not included any but the sparsest mention of Kuhn's biography. Philosophical biography is often interesting and sometimes illuminating, but rarely does it affect our assessment of the quality of the philosopher's arguments.

That said, I think it shows that Kuhn was never thoroughly trained as a philosopher. His undergraduate and graduate training were as a physicist, not as a philosopher, even if he described himself as a philosophically inclined physicist. Although he retired as a professor of philosophy at MIT, Kuhn was professor of the history of science at Berkeley when he published *The Structure of Scientific Revolutions* in 1962. Kuhn's treatment of philosophical ideas is neither systematic nor rigorous. He rarely engaged in the stock-in-trade of modern philosophers, the careful and precise analysis of the details of other philosophers' views, and when he did so the results were not encouraging (for example in his discussion of Kripke and Putnam on the causal theory of reference[3]). This is not to say that Kuhn was a bad philosopher or that we should be suspicious of his philosophical opinions – we might expect the grand, synoptic view to be characteristic of an important revolutionary thinker while the analysis of individual arguments might be cast as philosophy's parallel to "normal science". Even so, for a philosopher whose main achievement in the eyes of many is to have undermined a whole philosophical tradition (that of logical positivism, or more broadly, logical empiricism), it is perhaps surprising that he makes little direct reference to the claims of that tradition; even less does he give chapter and verse to establish that such-and-such is what the logical positivists did indeed think. This fact I think is further evidence that *The Structure of Scientific Revolutions* is primarily something other than philosophy – of the 150 footnotes in the first edition of that book only 13 include references to philosophers, and almost all of these are to philosophers whose views are in tune with Kuhn's (the vast bulk of the remaining references are to historians).[4] The importance of my remarks on Kuhn's relationship to the practice of philosophy is that the imprecise nature of the latter makes it much more difficult to assess exactly where Kuhn's differences with positivism and empiricism lie, how deep and extensive they were and how justified they should be judged to be. The usual assessment, especially in the light of Kuhn's huge impact, is that the break must be massive, a root and branch rejection, a thorough revolution. This I think is wrong. A central thesis of this book is that in important respects Kuhn failed to break entirely with the preceding tradition. From the naturalistic perspective that has developed in "core philosophy" during the last two to three decades, which in due course spread to the philosophy of science, Kuhn's views are shot through with commitments to the Cartesian and empiricist traditions he saw himself to be rejecting. Furthermore, I argue that it is the only *partial* rejection of positivism and empiricism that explains the radical appearance of the Kuhnian viewpoint – incommensurability, the conception of progress, the rejection of the concepts of truth and verisimilitude and, arguably the world change thesis, are all consequences of positivist and empiricist views that Kuhn retained. Had Kuhn gone the whole hog and *really* rejected empiricism then the result, although superficially less dramatic, would have been in fact a more truly profound revolution. In fact, as hinted, I think that many philosophers, including philosophers of science, now find themselves in the position of being rather less empiricist than Kuhn. At the same time, while philosophers of science have learned their lessons from Kuhn, in particular with regard to the importance of being historically realistic, the philosophy of science is today rather closer to core philosophy than at any time since the impact of *The Structure of Scientific Revolutions* made itself felt. In this context, a subsidiary aim of this book is to contribute to this reconciliation between philosophy of science and the rest of the philosophical family by showing, in a philosophical sense, how the rift occurred and how it may be healed.

The third, and I hope least important, limitation of this book concerns the huge secondary literature dealing with Kuhn's work. It has not been within my power to read more than a fraction of it, let alone discuss it in the text. There will be many commentators who in one way or another have helped form my ideas but who do not receive acknowledgment beyond the bibliography. From these I beg under-standing. There is however one book that I have found invaluable in my research and which deserves special mention. This is Paul Hoyningen-Huene's *Reconstructing Scientific Revolutions: Thomas S. Kuhn's Philosophy of Science* (1993). Hoyningen-Huene had the advantage of collaboration from Kuhn himself and his book is a scholarly and reliable guide through Kuhn's thought; it is essential reading for any serious student of Kuhn. The aim of Hoyningen-Huene's book is apparent from its title – it seeks to reconstruct Kuhn's philosophy of science. It does not especially seek to locate Kuhn in his intellectual context although the author is clearly aware of it – I aim to provide some of this. Even less does Hoyningen-Huene aim at a criticism of Kuhn's thinking, the scale and detail of the reconstruction providing a big enough task alone. Criticism is, however, the central purpose of my book. I should also like to thank Rupert Read and Wes Sharrock for allowing me to see a draft of their book *Kuhn: Philosopher of Scientific Revolutions* while I was in the final stages of writing this one. While there are several interesting similarities between their conclusions and mine, Read and Sharrock take overall a rather different approach. The first major difference is that where I see *The Structure of Scientific Revolutions* as presenting a historical theory that has important philosophical aspects, they see the book as more philosophical in conception, in a Wittgensteinian sense, providing a critical therapy of certain deep misconceptions in both the history and philosophy of science. This leads to the second important difference. We all see that there is residual empiricism in Kuhn's thinking, but because I see Kuhn's project as theoretical, I regard this remaining empiricism as undermining important aspects of it. On the other hand, since Read and Sharrock focus on Kuhn's critical intentions, they regard this empiricism as less serious. It may be that ultimately the areas of difference are not so important, and that what we are doing is illuminating different sides of the same coin, corresponding to the way in which, as I claim in my conclusion, Kuhn may be seen either as the last of the logical empiricists or the first of the post-logical empiricists. I look forward with interest to the completion of their book.

The first chapter of this book aims to introduce Kuhn's thought first by placing it in its historical context and then by giving a sketch of its main outlines. The following two chapters expand on the detail of Kuhn's theory: Chapter 2 explains Kuhn's belief in the cyclical nature of scientific change, analysing in detail his notions of normal science, discovery, crisis and scientific revolution; Chapter 3 describes Kuhn's explanatory mechanism, the theory of paradigms, exemplars and disciplinary matrices. These two chapters start with a detailed exposition of Kuhn's views followed by a critical discussion. My intention is that Chapters 1, 2 and 3 should be accessible to readers who have little or no experience in philosophy. My hope is that the same readers will also be able to follow the philosophically more intense chapters in the second half of the book, which concern Kuhn's more explicitly philosophical beliefs on the nature of: perception, observation and world change (Ch. 4); incommensurability and meaning (Ch. 5); progress, truth, knowledge and relativism (Ch. 6). Inevitably these chapters will be more easily accessible and initially, at least, more rewarding for readers with some philosophical background.

The bulk of this book was written with the benefits of a Leverhulme Research Fellowship – I am very grateful to the Fellowships Committee of the Leverhulme Trust and its secretary, Mrs Jean Cater, for their support for this project. The project got off the ground thanks to assistance from the Research Fund of the Faculty of Arts at the University of Edinburgh. My thanks go the Faculty and its Dean, Dr Frances Dow, for their early encouragement. Many friends and colleagues have commented on written drafts as well as providing helpful discussion in various forms; not all have agreed with my views but all have helped me form or reform them. In particular I should like to thank David Bloor, Arthur Campbell, Matthew Elton, Richard Foggo, Richard Holton, Peter Lipton, Rae Langton, Majeda Omar, Rupert Read, Michael Ruse, Lucy Wayman, Ralph Wedgwood, Michael Wheeler and Timothy Williamson.

I would also like to thank Cambridge University Press for permission to use the figures on p. 100, taken from N. R. Hanson's *Patterns of Discovery* (1965).

Chapter 1

Kuhn's context

Revolution in the history and philosophy of science

Thanks in large part to the work of Thomas Kuhn it has become something of a commonplace among historians and sociologists of science, and even among philosophers, that in order to appreciate fully the significance of a scientific idea as much as any other historical act, we must properly understand the historical context in which it arose. Kuhn's own work concerned only ideas in science, but soon after the publication of his book *The Structure of Scientific Revolutions* in 1962 his model of theory-change in science was being applied to the development of thought in a wide variety of areas of academic, intellectual, and social activity. While not all extensions of Kuhn's thought are equally sound, it would be odd if the injunction to consider the historical context were applicable to science but not to the history and philosophy of science. Accordingly, in this chapter I shall sketch an outline of some of the more important developments in the philosophical, social, and historical study of science in the years preceding Kuhn's work; a brief résumé of the central claims and concepts of Kuhn's thinking will conclude the chapter. An historical introduction is particularly apt and important in Kuhn's case. One reason for this lies in the very fact of Kuhn's basic ideas having been so hugely influential. As Kuhn himself argued, a new and revolutionary idea may come to form the basis of subsequent thinking in such a way that it becomes difficult to imagine not thinking in terms of that idea, and difficult therefore to appreciate the thinking that occurred before that idea came along. Consequently the requirement of sensitivity to historical circumstance is both less easy to achieve but also all the more important if we are to see why the new idea was so new and so influential. Since, necessarily, the originator of an idea started his thinking before the revolution his idea causes occurs, his perspective on its content and significance may well be different from ours – and there is reason to think this is true of Kuhn too. Some of his more fervent admirers and would-be disciples expressed thoughts or even found them in his writings that he had to repudiate as different from, or even repugnant to, his own thinking. And so, in order to be fair to Kuhn, we must be careful to read him as addressing a set of problems and concerns that he believed existed in the history and philosophy of science at the time when he was writing. These are unlikely to be the same as ours now. For that matter those concerns and problems did not remain constant throughout his life, and so we will need to give some thought to the changing nature and context of Kuhn's own work.

Because revolutionary thinkers are born and bred before the revolution they create, inevitably much of what such thinkers believe or, more especially, take for granted, will be shared with the thinking of those whose ideas they help to overthrow. In some regards revolutionaries will seem more advanced to their followers than they actually are. We are apt, for example, to think of Copernicus as the first modern cosmological thinker, the revolutionary who overthrew the Aristotelian-Ptolemaic picture of the Earth-centred universe. Yet, as Kuhn is at pains to illustrate in his first book *The Copernican Revolution* (1957), Copernicus may equally be regarded as one of the last Aristotelian cosmologists. While the eventual outcome of the revolution he ignited was the rejection of Aristotle's physics, there is no hint of this from Copernicus himself. On the contrary, Copernicus adhered to the Aristotelian premise that natural motion in the heavens is uniform and circular, and even claimed it as an advantage of his system over Ptolemy's in that it followed this precept more closely.[1] In many other respects, such as his belief in the finiteness of the universe, the existence of crystalline spheres, and the distinction between "natural" and "violent" motions, Copernicus was closer to Aristotle and Ptolemy than to us.

This provides the second reason why attention to Kuhn's historical circumstances is important. For, like Copernicus, Kuhn did not represent a total break with his predecessors. There are the seeds of Kuhn's own revolution in such historians and sociologists as Ludwik Fleck, Karl Mannheim and Robert Merton, as well as philosophers such as Toulmin and Hanson. More importantly, there are several significant respects in which Kuhn accepted the assumptions of those against whom he was reacting. Indeed, I shall argue that part of what makes Kuhn appear so radical is, odd though this may sound, precisely that he failed to reject much of the philosophy of science of his predecessors and contemporaries. The idea that revolutions exhibit some continuity with the past is perhaps well established; it is one to which we shall return in the scientific context. The analogy between scientific and political revolutions predates Kuhn.[2] The Russian Revolution was certainly a revolution, and a dramatic one at that. But both Tsarist and Bolshevik regimes used repressive techniques – secret police, spy networks, internal exile or labour camps – to maintain and exercise authority. Governments on neither side of the Revolution trusted the Russian people to control their governments democratically. Those few who advocated a constitutional monarchy or democratic republic along western European lines were seeking a revolution that would have been less dramatic but in reality more radical than that achieved by the Bolsheviks. Thus it was the similarities with Tsarism that made the Bolshevik revolution possible, and the dissimilarities that made liberal democratic institutions improbable. With no more analogy intended than this, I shall be arguing that likewise the dramatic aspects of Kuhn's own revolution are consequences not simply of what was new in his thought but in equal part attributable to what he retained. In this book I shall not only describe Kuhn's thought but also attempt a critical analysis. The outcome is that had Kuhn rejected *more* of his predecessors' thinking he might have come to conclusions less apparently dramatic than those he actually maintained – and also closer to the truth.

The Old Rationalism and the New Paradigm

It is now time to look at the world pre- and post-Kuhn – the world of the philosophy of science, that is. The name I shall give to the dominant network of ideas in pre-1960s philosophy of science is the "Old Rationalism", while its post-Kuhnian rival will be the "New Paradigm" in honour of Kuhn's most famous notion. The main players representing the Old Rationalism were Rudolph Carnap, Carl Hempel, Karl Popper and Imre Lakatos. These philosophers did not agree on all their opinions – quite the contrary. But they did share some views, both explicit and implicit. In particular, I use the term "Rationalism" to highlight the opinion common to all those

3

mentioned that a central task of the philosophy of science is to say what scientific rationality consists in, to describe how scientists ought to make inferences from evidence or to choose between competing hypotheses. This is *normative* philosophy of science. The rationality of scientific procedures, especially with regard to theory assessment and choice, may, as adherents of the Old Rationalism understood it, be encapsulated in the existence of the *scientific method*. Those adherents did not always call their accounts of scientific rationality by this name. But we may use it to capture a set of ideas to which they by and large subscribed. All Old Rationalists regarded the task of the philosophy of science as the articulation, in as much detail as possible, of the content of the scientific method.

One attempt to describe the scientific method, as here defined, was Rudolph Carnap's programme of inductive logic. Carnap hoped that philosophers could do for inductive reasoning something akin to what they had already done for deductive reasoning. As regards the latter, Frege and Russell laid down rules whereby the validity of supposedly deductive inferences could be assessed. Since inductive inferences are never valid in this sense, inductive logic would have to aim for something slightly different. Rather than establishing the truth of a conclusion given the truth of premises, as deductive logic does, Carnap's inductive logic aimed to establish the logical probability of truth of an inductive conclusion, given the evidence.

A quite different Old Rationalist proposal for the content of scientific method was Karl Popper's falsificationism. Popper fully accepted Hume's argument that inductive knowledge is impossible. The reliability of induction cannot be known *a priori* by reason alone. But any attempt to establish the reliability of induction on the basis of experience would beg the question of the reliability of induction, since it would inevitably employ induction. Thus Popper held that no rationally acceptable pattern of reasoning could establish the truth or even probability of an inductive hypothesis. Since he held that science is indeed rational, it follows that in Popper's view science does not in fact proceed by inductive inference. While a general proposition cannot be verified by corroborating evidence (results of experiment, observations etc.), such a proposition's falsity can be established deductively from refuting evidence. A single black swan (as may be observed in Australia) refutes the hypothesis that all swans are white. Popper built his picture of scientific method on the basis of the deductive validity of refutation. Science proceeds by a cycle of bold conjecture followed by the attempt to falsify that conjecture. Some theories will be refuted while others may survive (for a time). Following the falsification of a favoured hypothesis, scientists will try to come up with new conjectures and then set about attempting to falsify these. This process of conjecture and refutation Popper likened to Darwinian natural selection.[3]

Ultimately neither Carnap's logical probability nor Popper's falsificationism worked. Carnap's approach was limited in scope and in any case depended on the choice of language in which to express the hypotheses and evidence. It is clearly unsatisfactory to have the probability one attaches to a hypothesis dependent on the language one chooses to express it – unless there are rational grounds for preferring one language to another. As we shall see, Carnap thought this choice arbitrary. But if one thinks differently, that science should make our choice of language for us, then one faces a problem of circularity. For some language must be adopted in which to carry out science, but that will determine which scientific conclusions we may draw, including a decision about the language that should be in use.

On the other hand, Popper's avoidance of induction fared even worse. At best, according to Popper, we can only know that a hypothesis is false – we can never know that it is true. Despite Popper's attempts to suggest that we might get to know that one theory is *nearer* to the truth than another, even this cannot be done if, like Popper, we eschew induction altogether. This is because claims about the relative verisimilitude (truth-likeness) of general hypotheses always concern infinitely many consequences of whose truth or falsity we are ignorant. Consequently there is no deducing of facts about verisimilitude from our evidence; and so, according to a consistent anti-inductivist, there is no knowledge of verisimilitude. In fact things are worse than this, because we cannot even claim negative knowledge of the falsity of hypotheses. Popper accepts that observations are theory-dependent. An observation is theory-dependent (in one sense) if when making that observation we depend upon the same theory. For example, one might use a radar gun in order to measure the speed of a car or some other object. In so doing one will be depending on the theory of the Doppler effect in order to come to one's conclusions. A number of philosophers, Kuhn among them, argued that this theory-dependence is ubiquitous and Popper agreed. But if we do not *know* the theories in question to be true, as Popper holds, then he is committed to agreeing that we do not know the truth of any observation claims either. But observations are what falsify hypotheses. So, if we don't know any observations, we don't know that any hypotheses are falsified. And so Popper is committed to a very radical form of scepticism indeed.

It was partly this failure of the Old Rationalism to provide a satisfactory normative philosophy of science that gave rise to the "New Paradigm". One of the key representatives of the New Paradigm, Paul Feyerabend, challenged the notion of a method that would be applicable in *all* circumstances – the only injunction that could have such generality would be the empty rule "Anything goes". But if the Old Rationalists were right in thinking that a philosophically explicable scientific method or *a priori* account of scientific inference is an essential part of the picture of science as a rational enterprise, then the lack of such a method would seem to suggest that science is not so rational after all.

The Old Rationalists believed not only that they could give an account of scientific rationality but also that such an account could explain the actual development of science. They were impressed by the actual achievements of science and regarded scientific knowledge as the model for all knowledge. It is because scientists by and large do what a normative philosophy of science tells them to, or would tell them, that science converges on the truth and scientific knowledge accumulates. The Old Rationalism endorsed what might seem to be quite a natural picture of the history of science. We are inclined to think that we now know many scientific facts that were not known 1,000 years ago, and indeed many that were not known 100 years or even a decade ago. Our daily newspapers have columns devoted to recent scientific discoveries; the burgeoning of high-tech industries and the ubiquity of their products are everyday, if indirect, testimony to the more basic science that underlies them. It is only natural that we should think of modern scientific knowledge as *cumulative*. As time passes, new discoveries are made and new knowledge added to the common stock.

Reflection and a little knowledge of the history of science may refine this picture. For it is clear that not everything scientists have believed is true. Sophisticated observers before Copernicus believed that the Earth stood at the centre of the universe, while now we know it to be rotating (more or less) about the Sun, which is itself moving through space. And more recently too, once seemingly well-established theories have come to be held in doubt. It was thought that dinosaurs were cold-blooded creatures whose demise is explained simply by the forces of natural selection favouring warm-blooded mammals; today it is widely believed, if not yet conclusively known, that they were warm-blooded creatures whose end was precipitated by the impact of a huge meteor. The simple cumulative picture of progress will have to be amended to allow for the existence of false theories. But it is no great modification to allow that along with the large bulk of true theories, some false theories will be allowed to prosper temporarily. Furthermore, the amended picture will add that part of the growth of knowledge will involve the replacement of false theories by true ones – or at least by ones that are nearer to the truth.

If this traditional, cumulative picture were correct, the Old Rationalism had a ready explanation of that fact. According to the Old Rationalism, the reason why the history of science is one of increasing knowledge and of the replacement of falsehood by truth is that scientific beliefs are formed and theories accepted for purely rational reasons. As time passes our stock of evidence grows. Since the likelihood of a theory being true depends on the strength of the supporting evidence, and since rational individuals proportion their beliefs to the evidence, it follows that the rationality of science will explain its accumulation of truth. And if such an explanation is not available to Popper, he was at least able to make use of the Darwinian analogy. If poor theories are weeded out by falsification, those that remain will be more sophisticated and less liable to refutation than their predecessors. Explaining events in the history of science, according to this view, is much more straightforward than explaining events in human history outside science. To explain a scientific discovery it is usually sufficient to detail previous discoveries and the relevant available evidence, along perhaps with a description of the train of reasoning leading from these to the rational scientific conclusion, the discovery in question. Not invoked are factors *external* to the rational gathering and assessment of evidence. Whereas in political history we must refer to the characters, ambitions, prejudices, alliances, organizations and powers of men and women, in science we can ignore such matters. Or at least these need only be referred to on those rare occasions when it is necessary to explain a wrong turning in the history of science.

The fact that the Old Rationalism was so readily linked with this picture of science as a rationally ordered exercise in the accumulation of knowledge was a second source of dissatisfaction. Despite its attractions, it is a view that has come under considerable attack. Feyerabend argued that scientific heroes, such as Galileo, advanced precisely by going against the proposed canons of normative philosophy of science. Historical studies have suggested that the influence of what I called "external" factors is far more pervasive than the traditional picture allows, and that they are to be found at work in successful, "correct" discoveries as well as erroneous science. Sociologists of science have argued that social relations between scientists are among the factors that determine which theories get accepted. Indeed, they claim that the very kinds of thought a scientist can have are culturally or socially determined.

The failure of the Old Rationalist normative philosophy of science, along with the rejection of the traditional history of science that accompanied it, marked the advent of the New Paradigm. According to the New Paradigm there is no single, universal, all-encompassing scientific method that explains scientific development. Since the scientific method or something like it was held to be the essence of scientific rationality, the absence of the scientific method suggests that scientific change is not driven by the rational choices of scientists. Instead it is to be explained by historically local, sociological factors.

If one thinks that scientific change is driven by sociology not by rationality then one is likely to doubt that such change converges on the objective truth. One might think that this would require a commitment to scepticism – if science doesn't get us close to the truth, it doesn't give us knowledge. In fact, the New Paradigm tended to embrace less scepticism than its cousin, relativism. In giving up on our access to objective truth and hence on objective knowledge, the New Paradigm replaced these with relative truth and relative knowledge. Here "relative" means "relative to some social environment". Thus some proposition might be "true" relative to one community, and "known" in that community, while "false" and "not known" in another community.

Later we shall consider whether this sort of relativism stands up to critical scrutiny. Kuhn's own view is sceptical and incorporates a mild form of relativism. Other adherents of the New Paradigm, including those who professed to be Kuhnians, adopted a more radical form of scepticism. If one allies the truism "it is a fact that *p* if and only if it is true that *p*" to a relativist view of truth as always truth relative to a community, then it follows that what facts there are is also relative to a community. (Remember that we are talking of all scientific facts, not just facts about the community.) If "reality" is just what facts there are, and facts are community-relative, then what "reality" is, is dependent on the community. This extreme relativism is one strand of "social constructivism", holding that reality is not something that exists by and large independently of what the community thinks of it, but is constituted by the community's beliefs. It must be emphasized that Kuhn repudiated this view, but it is nonetheless true that Kuhn's work certainly gave encouragement to it – and it will be interesting to see whether mild relativism and radical social constructivism can be kept apart.

Logical empiricism and logical positivism

I have said that in order to understand fully the content of Kuhn's thought it is necessary to see him not only as rejecting many of his predecessors' ideas but also as continuing to maintain others. We will need therefore to look in greater detail at the philosophical milieu from which Kuhn's thinking emanated. The dominant philosophy of the twentieth century until Kuhn's time was logical empiricism.[4] Empiricism, whose modern form originates in the seventeenth century with Thomas Hobbes and John Locke, is the doctrine that all substantive knowledge and all concepts are derived from experience. Empiricism has had a long influence on the philosophy of science, not least because its emphasis on experience as the foundation of knowledge is readily interpreted as emphasizing the basis of science in observation. Note that there are two aspects to empiricism. One says that substantive knowledge can be had only in virtue of experience of the world – an *epistemological* thesis. The other aspect says that our concepts and ideas originate with impressions given to use through our senses. This is a *semantic* thesis, concerning the content of thought and the meanings of words.

The most influential empiricist of the late nineteenth century was Ernst Mach. Mach's outlook was thoroughly positivist. Positivism, which originated with the French philosophers Henri de Saint-Simon and Auguste Comte, is marked by a rejection of metaphysics and an emphasis on scientific knowledge as the only genuine form of knowledge. For Comte and Mach, the rejection of metaphysics includes the avoidance of unobserved entities in scientific theorizing, and the eschewing of such notions as "cause" and "explanation". For these philosophers the point of science is to describe – but not to explain – phenomena in a general way that will allow for accurate predictions concerning the same phenomena.[5] As a matter of actual fact, scientists *do* use "models", theoretical constructs that employ unobserved causes. A good example is Dalton's atomic hypothesis, which explains the law of constant proportions in terms of unseen molecules made up of chemically indivisible atoms. Mach was willing to allow that such hypotheses might be fruitful. But, ultimately, all that was properly scientific about any theory consisted in what could be said in terms only of observables. Mach suggested that humans tend to "economy of thought" – ways of thinking that encapsulate or sum up our experiences in a manner conducive to correct predictions of our experiences. Sometimes hypotheses about unobservables contribute to such economy of thought. But their worth is only in their predictive success and optimally they should be replaced by laws concerning only the connections of observable phenomena.

In rejecting metaphysics Mach also rejected traditional philosophical disputes and problems as misconceived – they are pseudoproblems. One such concerned the apparent dichotomy between the mental and the physical. For Mach there is only one kind of stuff, which is neither physical nor mental but which is the foundation both for what we regard as mental and what we regard as physical. Mach's building blocks were sensations. So all there is in the world are sensations, and physical objects are just one sort of grouping of sensations, while minds are a different sort of grouping of sensations. In this Mach's *neutral monism* is close to its empiricist predecessors such as Berkeley's idealism and Mill's phenomenalism.

If sensations are all that there is and all that we may experience, then the semantic component of empiricism requires that all our concepts must be explicable in terms of sensations. This is the basis of the logical empiricist interest in linguistic analysis and the philosophy of language. The German mathematician and philosopher Gotlob Frege distinguished between the *sense* of an expression and its *reference*. The reference of an expression is the object it denotes; the sense is the linguistic meaning of the expression. The sense also fixes or determines reference. Thus the name "Venus" might have a sense like "second planet from the Sun" or "the brightest planet seen at dawn or dusk", while the reference is that particular planet. In Frege's view, an expression such as "the present King of France" would be an expression with a sense but no reference. According to Bertrand Russell the expression is misleading. It *looks* like a term that might refer, but in fact is not. In Russell's view the sentence "the present King of France is bald" really says "there is exactly one thing which is a present King of France, and that thing is bald". In the latter sentence, the expression "*the* present King of

France" has been eliminated, so the issue of whether or not it refers does not arise. As far as Russell was concerned, all sentences of a language could be analysed in this way, in principle, into a perfect language. In the perfect language there would be a basic stock of expressions that refer to things and properties; all sentences could be analyzed into sentences containing just these basic expressions plus logical terms. This view, shared also by the young Ludwig Wittgenstein, is *logical atomism*. For Russell, the basic expressions denoted the basic elements of experience (the logical atoms) – *sense data* – similar to Mach's sensations. The only expressions with reference are the basic expressions, while the sense of any sentence is given by its analysis into the perfect language.

The most influential philosopher of science in the middle part of the twentieth century was Rudolph Carnap. Carnap was born in Germany but spent the years 1926 to 1935 in Vienna where he was a leading light in the Vienna Circle.[6] He shared Mach's rejection of metaphysics, as did the other members of the Vienna Circle. Metaphysical problems would be dissolved by the analysis of language. In particular, Carnap eventually came to think differences between idealists, phenomenalists, materialists, and so on could be regarded as differences in their choice of different sorts of language, rather than differences of opinion on some substantive point. Nonetheless, Carnap had a preference for a phenomenalistic language, similar to Mach's approach, in which the basic expressions designated actual and possible sensations. This is because knowledge of sensations has the advantage of epistemological certainty.

Since the attempt to say something with metaphysical content is confused, such statements would be devoid of cognitive content, perhaps even meaningless. The only cognitive content a statement could have is scientific. This led, via suggestion of Wittgenstein's, to a criterion of meaningfulness – the verification principle. The verification principle states that the meaning of a sentence is to be identified with the manner in which it would, possibly, be verified. (Later Carnap weakened to requirement from that of verifiablity to one of confirmability or disconfirmability.) In the case of simple statements, in a phenomenalistic language, asserting the existence of a sensation at a particular time, this is straightforward. The subject knows directly whether or not the sensation is being had. Matters are less straightforward when it comes to the meanings of the theoretical statements of science. These would be verifiable if they were equivalent to some statement about sensations. This is the approach

11

Carnap adopted in his *Der logische Aufbau der Welt*.[7] In the manner of Russell, all theoretical statements have an analysis in terms of sensations; so, ultimately, the only expressions with any reference refer to sensations. Hence a statement using the word "atom" is not really about an invisible particle but is a statement about the sorts of sensation we have. No doubt the analyses would often be extremely complicated, but they nonetheless would have their meanings fully explicable in terms of sensation words.

Nonetheless, there is a problem with effecting this direct reduction of theoretical to observation (sensation) language. For insofar as there is a connection between theoretical statements and observations, that connection is a loose one. Consider two balls of the same size lying on a table, one made of steel and the other aluminium. They look identical and so the sensations being had in each case are the same. Their theoretical properties, however, are quite distinct; for instance their masses are different. Hence theoretical statements cannot be equivalent to statements about *actual* sensations. At best the difference could be identified with the sensations one *would* have, for example, if one were to pick up both balls.

This and other problems led to a weakening of the link between theoretical sentences and observational ones. The *double language model* divides the language of science into two – theoretical language and observation language. What distinguishes this view from Carnap's earlier view is that the terms or sentences of the theoretical language are not directly definable in the observation language. Even so, the two are linked less tightly by so-called "correspondence rules". These are generalizations containing both theoretical and observational terms. The idea is that these correspondence rules allow for the verification of theoretical statements via their links with observation statements. Thus they also allow the theoretical statements to get their meanings in an indirect manner from the meanings of observation statements. But instead of theoretical statements having a clear meaning expressible in observational terms, they now have their meanings is a less clear fashion. Carnap talks of the theoretical propositions and terms as forming a network that is "anchored to the solid ground of observable facts" by the correspondence rules.[8] The anchoring is both an epistemological and a semantic anchoring. As Carnap admits, unlike the earlier picture, this means that the meaning or interpretation of a theoretical expression is incomplete. But this, he says, does not matter, since the correspondence rules give us a sufficient understanding of the theoretical terms, where by "understanding" he means the ability to make observational predictions from the theoretical system. It should be emphasized that on this picture, while meaning is injected into the theoretical system via the correspondence rules, the axiomatic struc-ture of the theory plays a role in spreading this meaning about.[9]

A further development in this direction is *instrumentalism*. On this view it is a mistake to think of theoretical statements as being properly true or false. As Mach said, they are there only to help organize our observational data and give us observational predictions. So, if theoretical expressions cannot be defined directly in obser-vational terms, then we should not regard them as having meaning in any full-blown, proper sense. Nonetheless, it is true that we think of theoretical terms as having some sort of meaning. So in this view the meaning that a theoretical term has depends on its connections with other theoretical terms, while a theory as a whole can be regarded as having an injection of meaning through its connection with observa-tion statements, as on the double-language model.

Gestalt psychology

According to Mach, all perception could be understood in terms of independent units of sensation. So just as an oil painting is made up of many small brushstrokes, each one uniform in shade, we might think of the visual field of perception as being composed of many indi-vidual units of uniform sensation. Christoph von Ehrenfels, and later the Berlin school of "gestalt" psychologists led by Max Wertheimer, Wolfgang Köhler and Kurt Koffka, challenged this approach, prima-rily on the grounds that it failed to account for the phenomenal facts – the way perceiving seems to us to be. We do not experience the world as a myriad of independent units but see it instead as grouped into objects. Looking at a building we see it as a single entity, distinct from the background scenery; this contrasts with the intrinsic nature of the oil painting where two brushstrokes both representing the foreground are no more linked than one from the foreground and an adjacent one from the background. Mach claimed that our sense of the existence of a distinct object is due to additional sensations that arise when the eye traces the object's outline. In contrast, the psychologists mentioned sought to explain the principle of grouping partly in terms of irreducible relations among the parts – relations that can only be grasped by seeing the form, the *gestalt*, as a whole.

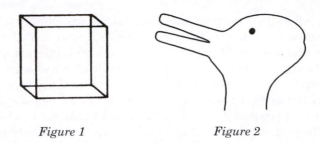

Figure 1   Figure 2

Various demonstrations exhibit our ability to see form in a non-atomistic way. Famous among these are the Necker cube (Fig. 1) and the duck-rabbit (Fig. 2). These show an ambiguity of gestalt – the same arrangement of sensational atoms can provide different experiences of form. The Necker cube can be seen as a depiction of a transparent cube or wire frame that comes out of the page towards the bottom right. But a shift in attention can cause the experience to flip to that of a cube coming out of the page to the top left. The duck-rabbit may be seen one moment as a duck with a beak then as a rabbit with ears. Implausible is an explanation of this in terms of a set of atomistic sensations plus the sensation of tracing the outline, since those would be the same for both of the two distinct gestalt perceptions of the duck and the rabbit, and of the cube in each of its two orientations.

The seeing of form and the ability to recognize objects, figures, melodies, and so forth are not aspects of perception that are additional to basic elements of experience, but are instead fundamental features of perceptual experience itself. Significantly, Wertheimer included past experience as one factor in fixing the form and structure that we perceive in the world (although he was careful to deny that this is merely a matter of repeated past associations). We shall see that Kuhn and others profitably exploited this idea.

Ludwik Fleck and the social study of science

Logical empiricism held sway in the philosophy of science, as well as many other branches of philosophy, throughout the 1920s and 1930s, and indeed, in several regards, until the 1960s. Nonetheless, just as empiricist ideas in psychology were criticized by the gestalt psychologists, non-empiricist ways of looking at scientific belief were also being developed. Largely neglected by philosophers when it was published, Ludwik Fleck's *Genesis and Development of a Scientific Fact*[10] provided a picture of the practices, concepts and beliefs of science quite at odds with the prevailing orthodoxy. That orthodoxy emphasized the centrality of the individual as the source and carrier of scientific knowledge. The foundations of such knowledge are located in the sense-experiences of the individual subject and are reported in essentially first person *protocol* sentences.[11] It is the subject who uses these protocol sentences as the foundations of a logically justified inference, the result of which is a state of knowledge in the subject.[12]

Fleck, by contrast, insisted that science and its knowledge cannot be understood in this individualistic fashion. The positivistic picture downplays the fact that scientific knowledge is the product of a collective enterprise. Individual scientist gets to know what they do know as a consequence of being taught or trained by others, by reading textbooks, and by being initiated in practical, experimental science; the scientist joins a research team, reads and contributes to journals and handbooks, and, not to be ignored according to Fleck, is a member of the general public, subject to its influences and participating in its general knowledge.

The positivist could not be accused of being ignorant of these facts of scientific life; but he would insist that they are irrelevant to epistemology. The distinction is made between the so-called "context of discovery" and the "context of justification". The former concerns the actual history of an individual or group coming to make some important discovery, while the latter refers to the relationship between a theory and the relevant evidence. So, for example, an account of the context of the discovery of the ring structure of benzene would include a mention of Kekulé's famous dream in which a snake bites its own tail. But neither Kekulé nor anyone else would regard the dream as a reason to believe in the ring structure. The reasons are to be found in the observed facts concerning benzene and whether the new hypothesis satisfactorily accounts for them. As Fleck himself puts it on behalf on the positivist, "it is not important to investigate how this connection was discovered, but only to legitimize it scientifically, prove it objectively, and construct it logically".[13]

While accepting the importance of "legitimization", Fleck regards the distinction between justification and discovery as too simplistic. Epistemology cannot ignore the historical process of discovery and hence cannot ignore the contribution of the social. To enable discussion of this, Fleck introduces his key notion of a *thought-collective*

15

(*Denkkollectiv*). This Fleck defines as "a community of persons mutually exchanging ideas or maintaining intellectual interaction".[14] As a definition this may seem somewhat loose. But even so it is not impotent, since important conditions have to be met before exchange of ideas and intellectual interaction are possible. To be able to exchange a scientific idea two persons must share the same scientific vocabulary, accept the same fundamental theories, recognize the same acknowledged facts; they must have similar notions of what a fruitful idea looks like and how it should fit appropriately with extant ideas; they must agree on the significance of the idea and on what would count as establishing it as an accepted fact. This set of shared beliefs and dispositions, Fleck names a "thought-style" (*Denkstil*).

The thought community cannot be reduced to a set of individuals, nor can the process of cognition – knowledge production – be characterized as a sequence of the thoughts of individuals. Despite the fact that much of the history of science is written in that manner, it gives us as realistic and informative a picture as, in Fleck's analogy, a report of a soccer game that lists the individual kicks one by one. Just as one must understand the play as the performance of a team as a unity trained for co-operation, knowledge production must be conceived of as an essentially collective activity. The knowledge of the thought-collective is more than any individual does or could possess, let alone be conscious of possessing.

Nonetheless, through the thought-style, collective knowledge will influence the thought and activities of individuals. If one ignores the thought-collective or denies its legitimacy as an epistemological concept, then one is liable to face other problems in the theory of knowledge, and "must introduce value judgments or dogmatic faith".[15] It is an undeniable fact that much of what an individual knows comes from or is dependent on the testimony of others; often the actual source of the testimony may be forgotten or unknown; it may be "common knowledge"; or the source may in turn rely on further testimony. What Fleck is driving at is that if we deny the legitimacy of the social source of much of even individual knowledge, we will find it difficult to give any justification to much of what we think we know. For example, the individualist about knowledge may seek to justify beliefs gained from testimony by reference to a justified belief in the reliability of the informant. That belief will be justified only if the subject is able to test a significant number of further reports from the informant against what the subject already knows. But, first, we rarely do test our informants' reliability to that degree; and, secondly, if we did, we would find that much of the knowledge being used to test the informant is itself gained from testimony. For example, if I wanted to check the reliability of a history book, I might check that it correctly records the dates of the battle of Hastings and the American Revolution and so on, dates I already know myself. But then the problem is that I only know these things because I read them in some other book or was taught them at school; that is, I would be checking the reliability of one piece of testimony by employing other pieces of testimony. The requirement that an individual be able to check the reliability of all testimony in order to have justified belief is an impossible one. Hence one may be left to a "blind faith" in the disposition of others to tell the truth. Fleck's alternative is to reject the individualistic conception of scientific knowledge.

Fleck's insistence on the irreducibility of the thought-collective, and on the mutual dependence between it and the thought of the individual, shows the clear (and acknowledged) influence of the father of modern sociology, Émile Durkheim. In developing his notion of a thought-style Fleck refers at length to the work of Durkheim's pupil, Lucien Lévy-Bruhl. What attracted Fleck was Lévy-Bruhl's emphasis on different kinds of mind or "mentality", exemplified by the extreme cases of the "primitive" mentality and the "civilized". The existence of distinct mentalities contrasts with the then prevailing view among British sociologists and anthropologists that there exists a single mentality common to primitive and civilized people. According to the latter view what distinguishes the two groups is a difference in belief. Primitive peoples have beliefs we can recognize as false; but their organization of experience in accordance with those beliefs obeys the same rules of reasoning as ours. Sir James Frazer, author of *The Golden Bough* (1890), argued that while humans have shown three stages in the development of thought – the mystical, the religious and the scientific – these nonetheless all reflect man's search for knowledge and desire for control of his environment. The evolution of thought comes about as a result of dissatisfaction with the failure of earlier beliefs to provide such knowledge and control. Thus, in general terms, Frazer regarded the thought of a man or a society as explicable in terms of universal reason plus fairly easily alterable beliefs. Lévy-Bruhl regarded mentalities as more deepseated than beliefs, but not universal either. Logic is a feature of civilized mentality that is absent from the primitive mentality, which, according to Lévy-Bruhl, is pre-logical.

Fleck's notion of a thought-style is an extension of the idea of a mentality. Different groups of "civilized" people may have different thought-styles. Furthermore, Lévy-Bruhl held that the civilized or scientific mentality involves a greater degree of sensitivity to "objective reality". Fleck denies that there is any such thing. What Lévy-Bruhl calls "proper" or "scientific" perception, Fleck regards as a learned disposition, acquired by induction into the thought-style of a particular collective. What Lévy-Bruhl thinks of as the civilized sensitivity to contradiction, for Fleck "is actually mere incongruence with our habitual thought-style".[16] The thought-style is thus key to determining our perceptions. But Fleck does not think that our perceptual beliefs are the products solely of our thought-styles. Two people looking down the same microscope at the same slide may give different reports, reflecting their different thought-styles. But those reports are not a function just of thought-style. If the slide is changed, there will be further, different, reports elicited. To mark the contribution of the thought-style, Fleck refers to "active links". The independent contribution is furnished through "passive links"; these we might think of as the contribution of some sort of reality.

Could we factor out the active links, removing the "subjective" element of perception? This Fleck denies.[17] He uses the following example to demonstrate the impossibility of pure objective perception. He considers an observer reporting what he sees in a bacteriological experiment. The supposedly pure assumption-free description might be: "Today one hundred large, yellowish, transparent and two smaller, lighter, more opaque colonies have appeared on the agar plate." Fleck points out that the statement implies that the two colonies constitute something different from the other hundred, and that they somehow belong together. This, he says, is not "pure observation but already a hypothesis which may or may not prove to be true".[18] Fleck goes on to point out that no two of the colonies will be identical in every respect. The degree of variation will render it impossible to record what is seen without *some* regard to what is an important difference and what is not. If one made an attempt to do so, by drawing up lists of the numbers of colonies with various characteristics, this would not permit new discoveries – that is achieved precisely by being disposed to notice significant differences. Nothing is "significant" in and of itself, but only in relation to some hypothesis or expectation. In any case, regarding all 102 objects as colonies itself manifests an assumption that none of these are non-bacterial "grains or dots".

Fleck linked his ideas on thought-styles to the gestalt psychologists' views on perception. The developed perceiving is made possible by a thought-style Fleck calls *Gestaltsehen* – the seeing of a gestalt.[19] Gestalts are "theoretically constructed" – Fleck says of one Joseph Löw, steeped in the thought-style of early nineteenth-century German *Naturphilosophie*, that he saw in the urine of infected patients a gestalt that we do not see. Fleck likened this to a case reported by the social psychologist Le Bon concerning the crew of a ship searching for a boat in distress. The sailors all "saw" the boat and people in it until they saw at the very last minute that it was a tree with branches and leaves.[20] Fleck's point is that not only pathological cases are like this – all are. Furthermore, scientific discovery is just like the sailor's changing gestalt. After the discovery "one can no longer even understand how the previous form [gestalt] was possible and how features contradicting it could have gone unnoticed".[21] In Fleck's view it is the thought-style that is responsible for *Gestaltsehen*, such as seeing the groupings that are the different individual and kinds of bacterial colonies.

Fleck was by no means the first to maintain that what we think and perceive is largely conditioned by social factors. But, as one of his early reviewers, Franz Fischer, pointed out, he was new in affirming that this applies specifically to science.[22] As we have seen, Lévy-Bruhl argued for a correlation between different ways of thinking and different levels of technical and social development (as had Auguste Comte before him). Karl Marx famously had argued for a *causal* connection, with the nature of a society's economic organization determining its prevailing ideology:

> The mode of production in material life determines the general character of the social, political, and spiritual processes of life. It is not the consciousness of men that determines their existence, but on the contrary their social existence determines their consciousness.[23]

Although Marx also thought that the rise of modern capitalism enabled Darwin to formulate the theory of evolution through natural selection, he tended to exclude a large range of thinking, scientific and mathematical thought in particular, from the general claim that the *contents* of ideas are determined by the economic and political interests of the ruling classes. The German social theorist, Karl Mannheim, writing in the 1920s and 1930s sought to generalize Marx's claim, both as regards the range of ideas and also concerning the variety of social factors that are relevant. First, there are more cognitive divisions in society than just class differences: other aspects of social status and role lead to the existence of distinct *Weltanschauungen* – world-views. Each world-view determined a style that is discernible in the products of the world-view, both intellectual and non-intellectual. Secondly, Mannheim criticized Marx for distinguishing between "bourgeois" and "revolutionary" ideas in the respect that only the former are socially distorted. According to Marx revolutionary thinkers are able to grasp the objective truth in a way that bourgeois thinkers are not. Mannheim rejected this epistemological privileging of one group. All groups and their ideologies are subject to sociological analysis: in every case it can be shown how thought matches social role. This led Mannheim's view to be characterized as relativist. But even Mannheim was reluctant to include the claims of the natural sciences among what he regards as "ideologies". The reviewer, Fischer, was right to point out the Mannheimian elements in Fleck's work and to regard as his original contribution as the extension of the *Weltanschauung* or thought-style idea to science.

Kuhn – the basic ideas

I have explained how the failings of Old Rationalist philosophy of science, along with the researches of Fleck and Mannheim among other sociologists and historians of science, encouraged the rejection of the picture of science as the accumulation of knowledge driven by a rational scientific method. Is there then a general description of scientific activity, of theory adoption and change in particular, that addresses the historical, sociological and philosophical concerns just mentioned? Thomas Kuhn's achievement in *The Structure of Scientific Revolutions* was to provide just this. So influential has that work been, that not only did it respond to this need in the disciplines of the history, sociology, and philosophy of science, but also more than any other work it has shaped the subsequent development of those disciplines.

Scientific progress, as seen from the traditional, cumulative perspective, is linear and uniform. There are no changes in the way that knowledge is generated nor are there profound changes in the kinds of knowledge produced. There is no change in the manner of scientific progress over time. The only change is in the quantity of scientific knowledge and this steadily increases. Kuhn's radical alternative rejects this. According to Kuhn we can identify distinct phases in the development of a science. In the different phases different kinds of scientific belief are generated, and generated in different ways. For a mature science the succession of these phases is cyclical – they follow a pattern that repeats over time. Furthermore, the nature of the cycle is such that we cannot straightforwardly say that scientific progress is cumulative. A later cycle does not simply take on board the discoveries of an earlier cycle and add to them, because important elements of the earlier cycle are rejected by the later one. I shall illustrate Kuhn's ideas using the history of astronomy, which was the subject of his first book *The Copernican Revolution* (1957).

The first distinction Kuhn makes among phases in the history of a science separates an immature period in the development of a science from its becoming a mature science. Kuhn has little to say about immature science – the focus of his interest is on the cyclical nature of mature science. It is clear however that people have always had beliefs on subject matters we would call scientific – the stars, the composition of matter, the nature of living things. But at the same time it is also clear that there has not always existed what we could call an established practice of science with agreed ways of validating these beliefs and adding to them, with research programmes that develop over time. Many ancient and primitive societies have beliefs about the cosmos, often with cosmologies giving some sort of description of the heavens. Frequently they identified the fact that the planets are different from the stars. Sometimes they had methods for calculating the positions of certain more important stars or even planets. This is all immature science. It was only with the ancient Greeks that astronomy was born as a mature science. The Greeks asked questions about the heavens that could not be answered by direct observations or simple calculations, questions that needed sophisticated methods for their resolution (e.g. What is the radius of the Earth? How far away is the moon? Why is the motion of the planets not uniform?). Cosmology and astronomy came together to give qualitative explanations of the nature of heavenly motions as well as quantitative predictions of those motions. It became an aim of astronomers to improve the accuracy of their models, thus instituting the first research programme in astronomy, necessitating both theoretical improvements to the model and new techniques of measurement to ensure the accuracy of key parameters. Thereby did astronomy become a true science, the culmination of this early development being the *Almagest* written by Claudius Ptolemy in the middle of the second century AD.

The *Almagest* contains a fully worked out system of the heavens. It tells us that the fixed stars are located on a sphere that is the outermost of a series of concentric shell-like spheres, at the centre of which lies the Earth. The Earth is fixed while all the spheres rotate about her. The Sun, Moon and planets each have a particular sphere whose rotational motion carries them about the Earth. This explains the fact that the stars, Sun, Moon and planets all move across the sky. However, not all of these move with a simple, uniform motion. In particular the planets, while appearing for the most part to move across the sky in one direction, from time to time reverse their motion, going back on themselves for a while before resuming their onward path. This is known as the *retrograde* motion of the planets. In order to explain this retrograde motion Ptolemy invoked the idea of an epicycle. The planets move on small circles – the epicycles – centred on a point within the relevant moving sphere. Hence, although the rotation of the sphere gives the planet an overall forward motion, the rotation of the epicycle will, from time to time, give the planet a temporary reverse motion. This solved many of the problems of getting a quantitative fit between his theory and observations, but by no means all. In particular the motion of the moon is difficult to account for in terms of a uniformly rotating sphere. To deal with this problem Ptolemy introduced another device – the equant – a mathematical point in space. While the motion of the moon's sphere is not uniform about the Earth, it is uniform about the equant.

Why did Ptolemy not allow for non-uniform motion, or for motion in some non-circular form that would allow for a good fit with observation? He was constrained by the requirement that his system obey what were then accepted as the laws of physics. These had been laid down by Aristotle and included the claims that natural motion in the heavens is always circular and uniform, and that the Earth is at the centre of the universe. According to Aristotle, the laws of physics are different for heavenly bodies and for things on or near the Earth (things below the sphere of the moon). These things have a natural tendency not for circular motion but for straight line motion, directed towards the centre of the Earth, towards positions of natural rest.[24] Another difference between the heavens and the Earth is that motion in the former is always natural whereas on Earth motion is sometimes "violent" or unnatural. We can propel a stone upwards, against natural motion, although the rule that "what goes up must come down" expresses the Aristotelian conviction that natural motion is a norm that will ultimately prevail. The possibility of violent motion on the Earth indicates yet another difference – the heavens are perfect while the Earth is corrupt. (In theological terms this is translated into the idea that Heaven is in or beyond the heavens (the planets and stars) while Hell is beneath us at the Earth's centre.)

Although the *Almagest* was extremely authoritative, it was not the last word. Medieval astronomers of the Arab and Christian worlds were aware that its quantitative fit was not perfect in every respect. Therefore astronomers undertook to see whether they could improve on Ptolemy's achievement. In the simplest cases improvements were sought only to the numerical values of certain parameters, keeping all of Ptolemy's structure. Even this prompted the making of new observations, such as were carried out at what was perhaps the first research institution of a kind that would be familiar to us – the Maragha observatory in Persia. This in turn required the design and construction of ever larger and more sophisticated instruments. Improvements in astronomical knowledge were motivated by, and applied in, the great church project of reforming the Julian calendar. At a more theoretical level, mathematical astronomers were developing geometrical techniques and devices for calculation. Others proposed additions to Aristotelian-Ptolemaic physics, such as impetus theory, or even suggested small-scale adjustments to Ptolemy's arrangement of epicycles, equants, and so on. All such developments were constrained and guided by the *Almagest*. All these researchers maintained the same format of an Earth-centred universe with planets moving on epicycles on spheres around the Earth. All adhered to the underlying Aristotelian physics. The geometrical complexity of Ptolemaic cosmology leads to considerable mathematical problems. Ptolemy himself devised the basic mathematical tools to solve these problems. The same tools and techniques were employed by his successors in solving their problems when constructing modifications to his system. Lastly, the *Almagest* set a standard of accuracy that acted as a minimum for any successor system.

The role played in medieval astronomy by the *Almagest* and its Aristotelian metaphysics illustrates Kuhn's notion of a *paradigm*.[25] A paradigm both guides and constrains research. On the one hand it sets the research problems. In the case of the *Almagest* the problem is one of finding ever-improved predictions of planetary motion. The *Almagest* guided research by providing the basic concepts and tools with which to do this – the ideas of the epicycle and equant, and the various mathematical techniques Ptolemy had developed. The problems Ptolemy himself dealt with and overcame provided examples for his successors of how scientific problems should properly be addressed and how they might be solved. On the other hand, a paradigm also provides constraints and limits on solutions to research problems. The *Almagest* sets a minimum level of accuracy, but it also sets a qualitative standard. Deviations from Ptolemy's system are permitted, but an excessive divergence from his system would be a sign of a poor solution.[26] In particular any satisfactory system would have to obey Aristotle's physics, by showing that the motion of the planets is circular and centred on the Earth. The structure of Ptolemy's system constrains research while improving on the system's details was the aim of that research.

Science that is governed by a paradigm Kuhn calls *normal science*. It consists in the search for solutions to problems set by the paradigm within a framework laid down by that paradigm. Kuhn calls this "puzzle-solving". Normal science is not the only state in which science may find itself, and a state of normal science will not last indefinitely. Practitioners of normal science may discover that certain problems are intractable and resist solution within the paradigm. In the course of normal science new phenomena may be discovered that cannot be explained using the resources of the paradigm. Such problems and phenomena are *anomalies*. In the case of Ptolemy's *Almagest*, astronomers in the fifteenth century became more fully aware of the failure of fit between that system and observation. Demands from navigators for more accurate techniques of prediction and from the church for calendar reform made this acknowledgment more acute. Neo-platonism also focused thinking on the mathematical elegance of the system – and its absence. Ptolemy's equant must always have seemed a somewhat ad hoc device, a device whose only role was to get the mathematics to come out right. It was clear to Copernicus and his contemporaries that a system without the equant would be a vast improvement over Ptolemy's.

The presence of anomalies is a fact of life for normal science. But an accumulation of anomalies that resist the best efforts to solve them may lead the science into its next phase, one of *crisis*. A science is in crisis when its practitioners are no longer convinced that the current paradigm has the resources to allow for the resolution of the mounting tide of anomalies. During normal science the inability of a scientist to solve a particular problem will reflect primarily on the capacities of that scientist. More charitably, failure may be put down to the fact that within the paradigm relevant facts have not yet been unearthed or appropriate techniques developed. But during crisis the failure to resolve anomalies will be blamed on inadequacies in the paradigm itself. Early sixteenth-century cosmology was in a state of incipient crisis.[27]

A science in crisis is unstable. If the central theory and paradigm of which it is a part are in serious doubt, then the paradigm will no longer be a suitable vehicle for guiding further research. A new paradigm is needed, one not beset in the same way by serious and intractable anomalies. The replacing of one paradigm by another is a *scientific revolution*. The term "revolution" is apt in two respects. First, it reflects the cyclical nature of change in a mature science. The adoption of a new paradigm as a result of a scientific revolution inaugurates a new period of normal science. The cycle – normal science, crisis, revolution, new paradigm, normal science – is complete. Secondly, there is an intended analogy between scientific revolutions and political revolutions.

In a stable society there are established mechanisms for resolving social and political conflict, such as elections and parliamentary debate, or even the will of a powerful autocrat. By whatever means laws will be made and will be recognized as laws. These mechanisms may be laid down in a written constitution, as in the United States, or may in part be given to us by established and accepted practice, as in Britain's unwritten constitution. The life of a stable society is analogous to the practice of normal science under a paradigm. It may be however that political conflicts arise that cannot be resolved in the normal fashion. Should there be sufficient of these, political and social tensions may grow to such an extent that they can be addressed not by change within the existing system but only by change of the system, just as science in crisis requires a change of paradigm.

In a normal political conflict all concerned are agreed what the mechanisms of resolution are and what counts as a satisfactory outcome. Thus a government may appeal to the support of the people in an election, and if it is defeated it will accept the result, passing the reins of power to some other party, as required by the constitution. But if these mechanisms fail to answer the problem in question, they will themselves be regarded as part of the problem. A revolution does not operate within a constitution – it seeks to overthrow the constitution. Therefore when a new political order and a new constitution are being sought, the old way of doing things and the old constitution will provide no way of determining which new order should be put in its place and will provide it with no legitimacy. A revolution is necessarily unconstitutional. Hence supporters of a particular new system will have to resort to other means to ensure its victory of rival proposals. They will have to resort to force and propaganda.

Similarly the transition to a new paradigm cannot be constrained by the old paradigm in the way that normal science used to be constrained by it. Crisis has discredited the old paradigm as a model for new theoretical development. The analogy with political revolutions has tempted commentators into thinking of a scientific revolution as a thorough and radical break with the past whose outcome is determined by highly contingent factors – the equivalent of the revolutionary mob, propaganda, coercion and so on. Correspondingly, they think that while in normal science what counts as a rational preference among competing theories is determined by reference to the paradigm, in the revolution the absence of a paradigm means there will be no agreed standard of rationality – revolutions are irrational. They pick up on Kuhn's remark that "external" factors – developments outside the science in question, and "idiosyncrasies of autobiography and personality"[28] – may play a major part in determining the nature of the new paradigm. "Even the nationality or the prior reputation of the innovator and his teachers can sometimes play a significant role."

This is one of the features of Kuhn's thought that has attracted most controversy. On the one hand it has led detractors to accuse him of making scientific development an exercise in irrationality, while on the other it has acted as a spur to the sociology of science, both in its micro form, investigating the way, for example, attitudes of the directors of research centres and chairmen of grant-awarding bodies influence paradigm acceptance, and its macro form, looking at the political, economic, and social determinants of the content of science.

In fact, as we shall see, Kuhn is not nearly as radical as this suggests, and he holds the major factor in paradigm choice to be the ability of a new paradigm to maintain the successes of its predecessor and solve its unsolved problems, a factor that is internal to the concerns of science. As I mentioned at the beginning of this chapter there is more continuity in even political revolutions than might at first be apparent. The same certainly goes for scientific ones. There is another feature of Kuhn's thought that tends to attract to it the charge of irrationalism. Since the paradigm not only supplies the framework for the development of puzzle-solutions but also the standard by which they are judged, once a paradigm has been overthrown and the rival to replace it are in competition, our usual standard for scientific evaluation is unavailable. There is no universal or common measure of theories – in the jargon, they are *incommensurable*. Kuhn was particularly struck by the difficulty presented to historians in trying to understand earlier theories. In his own case he found that at first Aristotelian physics seemed incoherent and full of obvious errors. But on further investigation it became apparent that Aristotelian terminology that appears to have a natural modern equivalent in fact means something quite different from its counterpart. A fuller understanding of the theoretical context enables one to have an understanding of Aristotelianism that no longer makes it appear straightforwardly wrong. Nonetheless, that understanding is not one that permits a new, improved, and exact translation between the vocabulary of Aristotelianism and modern physics. There is no common language that can express both theories.[29] The shift the historian undergoes may be described as learning to see the world in a different way. The same psychological leap occurs when scientists make the transition from one paradigm to another; they too will see the world in different ways. Just as Fleck said that operating within a thought-style is a matter of *gestaltsehen* – seeing a gestalt – Kuhn likens the experience of changing a paradigm to a gestalt-shift, where looking at the same figure the subject first sees one form then another, a duck then a rabbit, or a cube first from above then from below.

Since one paradigm does not build upon the achievements of its predecessors but instead overthrows them, the development of science cannot, in Kuhn's model, be a matter of the accumulation of truth. Nor, for similar reasons, does the history of science record ever improving nearness to the truth. But Kuhn does not reject the idea of progress altogether; he replaces these traditional accounts with an evolutionary picture. Science does not progress towards some goal (truth), but instead progresses away from its primitive, earlier stages. Kuhn thinks not only that truth and truth-likeness are irrelevant to the description of scientific progress, he thinks also that they are irrelevant to explaining it. This is what I call Kuhn's "neutralism" about truth. At one point Kuhn goes further and says that truth as traditionally conceived, that is as a matching between a theory and reality, is incoherent.

This completes my brief survey of Kuhn's views and their background. The remaining chapters of this book will both present Kuhn's ideas in greater detail and also subject them to critical scrutiny. In Chapter 2 I will look at the cyclical pattern Kuhn discerns in the history of science, while in Chapter 3 Kuhn's explanation of this pattern and its central component, the paradigm concept, are examined. Chapter 4 is concerned with the claim that perception and observation are theory-dependent and seeks to understand what Kuhn means by suggesting that when paradigms change, the world changes too. The incommensurability of theories and the meanings of scientific terminology are the subject matter of Chapter 5. Chapter 6 asks what kind of relativist, if any, Kuhn was. I conclude with a short chapter in which I first discuss Kuhn's influence on social science and then present an overview of my picture of his thought and its relationship to contemporary philosophy of science.

Chapter 2

Normal and revolutionary science

Theoretical history of science

The Structure of Scientific Revolutions is a work in the theoretical history of science. It is not philosophy, although Kuhn's theoretical standpoint meant that he had to engage deeply with philosophy. Nor, unlike his first book *The Copernican Revolution*, is it straight history of science, detailing and explaining a particular past episode or scientific development. Kuhn's treatment of the history of science in *The Structure of Scientific Revolutions* can be regarded as having two aspects. The first we may call *descriptive* – he details what he sees as a pattern or regularity in the development of the various sciences – the cycle of normal science, crisis and revolution. The second aspect is the *explanatory* side in which he tries to find an underlying explanation, some general feature of science that accounts for the pattern – this is Kuhn's theory of paradigms. In this chapter I shall discuss the descriptive aspect of his project, and in the next I shall examine the explanatory claim.

Kuhn himself did not clearly distinguish between the two elements of his project, instead combining the two as a single picture of science and its workings. This is one of the reasons for the vagueness and multiplicity of uses some of his critics found in his employment of the paradigm concept.[1] Since this theoretical, explanatory concept is frequently used in his descriptions of scientific change, it can be difficult to assess his descriptive claims. For example, if we define normal science as "a period of scientific research governed by a paradigm", and define revolutionary science as "research that is not normal", then it becomes tautologous to explain scientific revolutions as the overthrowing of paradigms. We also lack any criteria for distinguishing normal from revolutionary science. This of course is a caricature of Kuhn's views, but it does illustrate the obstacles in the way of a critical assessment of Kuhn's achievement. We must be grateful that he gives a sufficient wealth of detail and example to allow us to make a good fist of reconstructing the distinct descriptive and explanatory parts of his theory. The reason for starting with the descriptive aspect is simple – if the descriptive claims are badly mistaken, then the explanatory claims are that much less likely to be true. In fact we shall see that phases in the history of science are not quite so clearly differentiated in the manner that my brief description of Kuhn's account in Chapter 1 suggests, and that some important episodes seem to have little place in it.

Immature science

Before getting to grips with the various phases of the cycle of a mature science it will be helpful to have a look at thought that has not reached this stage of development. Because of the close relationship between normal science and paradigms, it is useful to start by thinking of this earlier, immature stage as being science without a single governing paradigm.[2] Without such a paradigm there is no consensus, no set of generally accepted facts, methods, examples, or research problems. This means that each researcher or group of researchers may start from scratch – and may generate new sets of data and employ new methods for analyzing the data or extending them. Correspondingly there is opportunity for a proliferation of approaches to the subject matter, and indeed there may not be agreement on what the subject matter actually is since there may be no accepted set of data requiring explanation.

This proliferation will be reflected in the existence of different competing schools of thought. Kuhn says that optical theory before Newton was like this. Different groups espoused views of light as particles emitted from bodies, as modifications of an intervening medium, and as an interaction between the medium and the eye. There were other groups besides, as well as subgroups supporting variants on these ideas. Since there were these competing groups, scientific activity aimed as much at refuting or convincing the competition as it did at pushing forward a program of research.[3]

At the same time it should not be thought that immature science is merely a fruitless bickering. There could be useful observation, experimentation, theorizing and conceptual innovation in the absence of universal consensus. (Although without that consensus such advances will not be universally acknowledged as such.) Eighteenth-century electrical researchers did engage in experiments that advanced understanding, and even the theorizing of ultimately unsuccessful schools had a positive influence on those ideas such as Benjamin Franklin's that did survive to become the first paradigm.

If the adherents of a school are genuine scientists, as Kuhn emphasizes they are, and if they do make advances based on the assumptions of their school, we may ask, together with Paul Hoyningen-Huene,[4] what distinguishes science carried out within a school from normal science under a paradigm? So far it looks as if the prime difference is just that in normal science there is a single paradigm while in immature science there are several competing paradigm-like schools.

The very fact of competition between schools itself makes a difference. As already remarked, competition means that effort is put into combating opponents rather than into cumulative research. Given the existence of alternative approaches and methods, a scientist must always clarify and justify his own approach and methodology rather than taking them for granted as in normal science. Consequently scientists cover the same ground over and over again and are able to make much less progress than in a state of universal consensus.

If competition is the characteristic feature of immature science, then we might liken that phase of science not to normal science but to revolutionary science. The revolutionary phase is marked by the existence of competing proposals for new paradigms, whose proponents are forced to put their efforts into explaining and justifying the foundations of their positions. Let us look then at another question that Hoyningen-Huene asks of the immature period: How does it come to be replaced by consensual normal science?[5]

The answer is that one school scores a victory over its competitors by producing a signal achievement. That achievement wins defectors from other schools and attracts the favour of younger scientists. Support for the competitors dwindles and they eventually die out. The achievement resolves a significant measure of the disagreement among the schools and provides firm foundations for future research:

"confidence that they were on the right track encouraged [scientists] to undertake more precise, esoteric and consuming sorts of work".[6]

This description of the transition from immature to normal science looks just like the resolution of a period of revolutionary science by the adoption of a new paradigm. And this should not be surprising since the result – a period of normal science governed by a paradigm – is the same in both cases. And so what distinguishes immature from revolutionary science must be not what happens during these phases but what precedes them and brings them about. Revolutionary science is precipitated by a crisis in an existing paradigm. But the immature science of competing schools is preceded by no paradigm at all, perhaps no science at all, for the phenomena of interest may previously have been unknown or ignored.[7] What initiates a period of immature science may be curiosity in a new phenomenon or a conviction that a certain field might reward study, just as hopes of transmutation, elixirs and the like may have stimulated alchemy. Pre-science may continue for centuries.[8] Since revolutions occur in mature sciences, those engaged in them are likely to have deep-seated, often professional interests at stake. The sense of crisis may impart an urgency to the search for a new paradigm. Revolutions will typically be brief affairs.[9]

This suggests a characterization of immature science. Immature science is that initial period in the history of a science that occurs when for the first time sufficient interest in a phenomenon or set of related phenomena crystallizes distinct groups or schools around particular theories or approaches; the schools not only pursue research on the basis of their favoured ideas but also compete with one another for intellectual, social and professional supremacy.

We have already seen what brings about a period of mature normal science. The first such period in the history of a field will occur when one school of immature thought vanquishes its opponents with some special achievement that forms the basis of the consensus governing the research that follows. Subsequent periods of normal science are instituted in an analogous fashion, following a revolutionary period, when one proto-paradigm succeeds in attracting the support of scientists in preference to its competitors. In both kinds of case the characteristic of the new period of normal science is widespread agreement on a core theory and basic techniques and approaches that must be employed in any satisfactory work in the field.

Normal science

In Kuhn's cyclical view of scientific change there are two sorts of scientific research that can take place when a field has reached maturity – normal science and revolutionary science. As the term "normal" science suggests, Kuhn thinks that most science is of this kind and not of the revolutionary sort. It is appropriate then that despite the title, *The Structure of Scientific Revolutions*, half of that book does not concern revolutionary science at all. In summing up the nature of normal science Kuhn says that during such a phase "scientists, given a paradigm, strive with all their might and skill to bring it into closer and closer agreement with nature".[10] When it comes to detailing what goes on in normal science, Kuhn is clear that more than this is involved. As regards both fact-gathering and theorizing, the activities of normal science fall into three classes: (a) determination of significant fact; (b) matching of facts with theory; (c) articulation of theory.

Only (b) directly fulfils Kuhn's general characterization of normal science as the co-ordinating of paradigm and nature; (c) fulfils it indirectly and (a), although it assumes and builds upon the paradigm, does not contribute to bringing it and nature closer.

Theories will often tell us, in general terms, that facts of a certain kind exist and it will then be natural to go out and collect and record such facts. This is what Kuhn means by (a): "determination of significant fact". In the context of the atomic theory of matter it makes sense to try to find out atomic masses and molecular structures; if the atomic theory were false there would be no such facts to look for. More recently, the human genome project that seeks to record all the loci on a typical set of human chromosomes makes sense only in the context of a theory that says that such things exist and have scientific significance. To this gathering of significant fact there is a theoretical counterpart, which is the making of predictions of such facts on the basis of theory. For example, given the theory of refraction it will be possible to calculate the optical properties of lenses of various shapes. Kuhn says little about this, remarking that scientists regard such problem as "hack work to be relegated to engineers or technicians".[11]

More significant are activities under (b): "matching of facts with theory". Such matching is not automatic – again there are related theoretical and fact-gathering activities. In establishing his paradigm Newton showed, using the mathematical techniques he himself developed, how Kepler's laws of planetary motion might be derived, subject to certain approximations, from his own laws of motion and gravitation. But getting a fit between his theory and the observed motion of the moon was a theoretical problem that Newton only partly solved and its full solution required the development of yet more sophisticated mathematical tools by his successors. In addition to such theoretical advances, improving the closeness of fit requires establishing certain important facts – some newly known facts and some that are more precise replacements for old ones.

In addition to improving the precision of fit in areas already addressed by core theory, theoretical and observational progress is required by the application of the theory to new areas outside the original field of application. Newton's laws were developed primarily with celestial mechanics in mind, and his applications of them to terrestrial problems were limited primarily to the problem of tides and to simple pendulums. A challenge then is to develop further applications to other terrestrial phenomena, such as compound pendulums, vibrating strings and hydrodynamics. Successful applications improve the fit between theory by extending the areas of contact between theory and observed fact. Theoretical and observational progress is required here just as in the case of improving fit in areas of existing contact.

The third class of research problems for natural science, (c), concerns the "articulation of theory". Here we need to treat fact-gathering and theorizing separately, since in this case they are not related.

Kuhn notes that a repeated and important theoretical task is the clarification of current theory by its reformulation. The earliest version of a theory, that given to us by its creator, may suffer from being a first attempt and from some of its ideas being only implicit in its applications. One might add that formal, mathematical techniques might later be developed that facilitate a different, more convenient formulation of the theory. Consequently Newton's theory in the hands of Hertz looks very different from what we find in *Principia Mathematica* although logically equivalent to it. A slightly different instance of this kind of theorizing would be the proof that the matrix mechanics and wave mechanics versions of quantum mechanics are mathematically interchangeable.

Clearly the formal recasting of a theory does not require any new empirical knowledge. And so there is no fact-gathering activity linked to this sort of theorizing in normal science. Nonetheless, Kuhn does say that there is another aspect of theory articulation that involves an intimate interaction of theoretical and empirical research.

Let us distinguish between theory application and theory extension. The former involves using a theory, perhaps in conjunction with new formal techniques and factual discoveries, to get a fit between the theory and nature. Although there is theoretical work it does not involve augmenting the core theory. Such problems of theory application fall into the second class, (b), of normal science activities. By contrast we might add new theoretical components to an existing theory. Newton's theories say nothing about electricity or magnetism. Thus Coulomb's law of electrostatic attraction is clearly an addition to Newton's laws. But it is an extension rather than a completely independent theory since it integrates with Newton's laws – the laws of motion along with Coulomb's law are required in order to predict or explain the motion of charged particles. Indeed, Coulomb's law is the perfect analogue of Newton's law of gravitation. Since this an extension of theory the gathering of new facts will be essential to providing empirical confirmation. In particular this will be required when there are competing possible extensions of theory.

In another of Kuhn's examples caloric theory was developed to account for the heating and cooling of substances caused by mixing them or by evaporation, condensation and other changes of state. To be a general theory of heat the caloric account had to be applied to other instances of heat production or absorption, such as chemical reaction or friction. But the extension of the theory to these other cases was not straightforward – the original theory could be developed in more than one way to account for them. Nonetheless the potential for development is not open-ended; it is constrained by the original theory. Much important work in this area is experimental, since it is carefully designed to decide between the competing theoretical extensions. I have already mentioned Coulomb's extension of Newton's laws to cover electrostatic attraction. The culmination of this line of extension was Maxwell's set of equations. These implied the existence of an aether. The famous Michelson-Morley experiment was designed to corroborate this extension, and in particular one version of the extension, a version that says the aether is fixed and unaffected by physical objects, in contrast to the aether-drift theory, which says that the aether may be dragged along by moving objects. In such cases the theoretical extensions and the experiments needed to test them are developed side by side.

There are other experimental activities that fall under the heading of theory articulation. One is the development of experimental laws, such as Boyle's and Hooke's laws, or more recently Hubble's law. These are not applications of existing theory, since such laws are experimentally discovered and not derived theoretically from existing theory, although that may happen subsequently. Nor are these laws extensions of existing theory since they do not deal with independent phenomena that could never be explained by the existing theory alone. Furthermore applications of experimental laws do not dovetail with the existing theory in quite the way that extensions do. Rather they are waiting to be explained by the existing theory, as Boyle's law was by the kinetic theory of gases, whose theoretical assumptions are purely Newtonian.[12]

A last fact-gathering aspect of theory articulation is the fixing of the values of physical constants, such as Newton's gravitational constant G, or Planck's constant h. This could be thought of as a special case of the gathering of facts significant according to the theory, because the existence of these constants is something asserted by the theory and is clearly significant by the lights of the theory. But since these constants and their values are so central to the theory, this kind of fact-gathering deserves separate mention. It is because physical constants are so special that experiments designed to measure them are frequently repeated with increasing precision, and new experiments are designed to confirm the results. The values of some physical constants are the most precisely known facts in science – the 17th General Congress on Weights and Measures adopted a value for the velocity of light that has eight decimal places; its value is thus known to a precision of fractions of a part per million.

It is a little surprising that Kuhn did not include among the theoretical work of normal science such theorizing as is needed in the course of fact-gathering, experimentation and observation. This frequently goes beyond "hack work" and can be highly sophisticated, on the same level as applications of theory. The first determination of the speed of light, by Rømer in 1676, required a careful application of Newton's laws to the moons of Jupiter. Thanks to the discrepancy between these predictions and what he observed, Rømer was able to calculate the difference in time it took light from Jupiter to reach the Earth, depending on whether Jupiter and the Earth were near or far apart. From this a (surprisingly good) value for the speed of light could be calculated. More recent methods, such as those based on the Err effect, are also theoretically non-trivial.

Kuhn is keen to emphasize that normal science, whether experimental or theoretical, is low on innovation and high on dogma. This must of course be implicitly a relative judgment. Although some of normal science involves the literal repetition of experiments performed before, even that is likely to be related to something new, such as the replication of a newly reported experimental effect. Most of what is described as research under normal science led to the discovery of new theories, new laws or new facts, even if only "old" facts known with new accuracy. So there must be an intended comparison next to which this research seems conservative. The comparison is, naturally, revolutionary research. The new in normal science is the embellishment of existing theory; in revolutionary science it is the replacement of existing theory. In normal science the theory is not up for debate. Research takes it for granted; it has therefore the status of dogma. Kuhn does not intend this characterization pejoratively. On the contrary, normal science could not progress without the unquestioned acceptance of a theoretical foundation, just as civil society could not function without constitutional consensus.

Paul Hoyningen-Huene says that the core theory is not tested, nor is it confirmed, by experimentation under normal science. Kuhn regards normal science not as testing the core theory but as testing the scientists. And so, says Hoyningen-Huene, experiments do not confirm the theory (even as an unintended consequence), because "confirmation, in the strict sense, can only occur against the background of possible refutation".[13] While Kuhn and Hoyningen-Huene are probably right as regards testing, the claim about confirmation does not follow as quickly as the latter seems to think. While the most powerful confirmations are those that might, beforehand, have turned out to be refutations, it is also true that perfectly good confirmations can occur without a corresponding possibility of falsification. Speaking generally, existential hypotheses can be confirmed by observations that would not have falsified those hypotheses had they turned out otherwise. For example the claim "there is a red kite in Suffolk" can be confirmed by a single lucky sighting, but the failure to see the bird on that occasion is no refutation. Similarly, Darwin's evolutionary hypotheses are confirmed by the fossil record, which shows signs of speciation (branching of species) and development within species. But the absence of such data for many animals (including the "missing link" in human ancestry) is not counter-evidence to Darwin. The incompleteness of the fossil record is hardly surprising, let alone damaging to Darwinism. Thus it may well be correct to say that success in matching theory and fact during normal science does confirm that theory. If you ask a scientist why we should think that a long-standing theory is true, you will find that the theory's historical successes are cited in its support. Scientists do not *think* that the answer is: the theory *cannot* be confirmed – it is a dogmatic foundation of our work. (Of course, they may be mistaken or deceiving themselves.)

It is nonetheless fair to say that scientists do not think of their normal scientific activity as testing the core theory, nor do they intend to confirm it. Outcomes that conflict with the theory are not considered as lying within the range of expected possible results. Novelties of that degree are not envisaged. The areas of normal science where there is most scope for innovation are those concerned with the discovery of new empirical laws. For there may not be any guidance contained within the core theory about which phenomena are nomically related, nor in what way. Newton's laws give no indication that the pressure and volume of a gas may be related, let alone by what function.[14] Even so, likely areas of fruitful research of this kind are typically suggested by observations and sometimes by theorizing carried out in pursuit of other kinds of normal research. In the context of the mechanical view of nature encouraged by Newton's laws of motion, it was natural for Robert Boyle to investigate the mechanical properties of gases. Hubble's law was discovered in the process of cataloguing galaxies and their characteristics. Other normal science activities – such as extending core theory, gathering significant facts or determining constants – are governed by the core theory and generate expectations about what sorts of result will be achieved.

It is because scientists have fairly clear expectations about the range or nature of possible results from experiments that they are able to design and build apparatus to give them useful results, often at enormous expense as in the case of particle accelerators. An experiment carried out speculatively, with no idea what results might be produced, will be next to useless since scientists will have little idea how such results relate to existing research.

Described thus, normal science might seem an intellectually uninteresting activity. Kuhn is at pains to stress that this would be misleading. The application of a paradigm to produce the facts mentioned above is not mechanical. It may require a great deal of ingenuity, mathematical competence and technical skill. Kuhn even suggests that it might require some conceptual innovation.[15] According to Kuhn, what makes normal science attractive is not the intrinsic, often limited interest of the experimental results or problem solutions; instead it is the intellectual challenge of the process of arriving at the solution. This Kuhn calls "puzzle-solving". Consider jigsaw and crossword puzzles, or chess and logic puzzles: the completed puzzle is itself an item of little worth. People enjoy such puzzles for the pleasure of the process of solving them and for the satisfaction, when successful, of proving their skill. Normal science is similar. The results will not be spectacular but the process of getting the results can be much to the credit of the successful scientist. Two further analogies with puzzle-solving are significant for illustrating the nature of normal science as governed by a paradigm. First, practitioners know that there is a solution. Just as someone having difficulty with a crossword clue doesn't doubt that there is an answer, the scientist knows that there exists a solution to his normal science problem. Failure to find the solution therefore reflects badly not on the paradigm but instead on the skill or ingenuity of the scientist. Secondly, not just anything will count as a solution. Jigsaw pieces that don't fit easily may not be forced together; the resulting picture must resemble that on the box. One cannot just put any old words into a crossword – they must relate appropriately to the clues in ways that the aficionado will understand but the neophyte may not. Similarly, the paradigm provides standards by which puzzle-solutions will be judged. Only certain kinds of result will be regarded as genuine solutions. James Watson's description of his discovery, with Francis Crick, of the structure of DNA illustrates Kuhn's idea. They both knew as did their competitor, Linus Pauling, and everyone else that DNA must have some chemical structure, and they all had a rough idea of what the answer must look like (some combination of molecular chains). The puzzle was to work out how many chains and which constituents. Part of the process of finding the solution involved picking up more "clues", which was not always done by them; one of the most significant clues was Rosalind Franklin's diffraction pictures of DNA. Some of their activities were more literally like puzzle-solving – building models with model atoms and molecules. One attempted solution was rejected since it involved stretching the bonds and their angles too much. At another point Watson makes progress towards the answer by trying to piece together molecules represented by appropriately shaped pieces of card.

Anomalies and discovery

The fact that normal scientific research generates expectations that are usually satisfied does not mean that such expectations cannot be disappointed. Returning to the case of the discovery of Hubble's law, Edwin Hubble was cataloguing galaxies as a result of his previous discovery that not all nebulae are part of the Milky Way; some are distant, independent galaxies. This discovery came about as a result of his work on Cepheid variables, a certain kind of star, for which a correlation exists between periodicity and (absolute) magnitude. In recording their observed magnitudes, which depend on their distance and absolute magnitude, it became apparent that some must have distances that exceed contemporary expectations about the size and nature of the cosmos. This exemplifies Kuhn's use of the term "discovery": "Discovery commences with the awareness of anomaly, i.e., with the recognition that nature has somehow violated the para-digm – induced expectations that govern normal science. It then continues with a more or less extended exploration of the area of the anomaly. And it closes only when the paradigm theory has been adjusted so that the anomalous has become the expected".[16]

Classic cases of anomaly are those where nature is in direct conflict with the core theory. But as we shall see, Kuhn regards paradigms as more than just theories – they can also include commitments to instru-mental procedures. Consequently Kuhn's examples of discovery include not only the discovery of oxygen by Lavoisier, which arose from his awareness of experimental results in conflict with the prevailing phlogiston theory, but also Roentgen's discovery of X-rays. Roentgen's observations of an unexpected glow on a screen at a distance from a shielded cathode ray tube led him to propose the existence of rays distinct from the cathode rays themselves but with some similarity to light. There was nothing in this that conflicted with existing theory. Nonetheless Roentgen's claim caused great controversy. According to Kuhn this was because it violated not theoretical expectations but experimental expectations associated with the widespread use of the cathode ray tube. If there was more going on in the cathode ray tube than its designers and users had known, then inferences from experi-ments of which it was a central part were correspondingly less reliable. We may call discoveries like those just discussed "anomaly-driven" discoveries, and if we want to maintain use of the word "discovery" for events such as the discovery of new elements or new quantitative laws, then we can call these "cumulative" discoveries. Cumulative discoveries simply add to existing beliefs, while anomaly-driven discoveries have a destructive element as well. They require giving up some previously held belief or procedure.

We have not yet discussed in detail what paradigms are, but it is clear that anomaly-driven discoveries typically require some revision to a paradigm; they result in paradigm shifts. Do they thereby constitute revolutions? For Kuhn, revision of existing belief is a necessary feature of a scientific revolution, but is it a sufficient condition too? The standard and straightforward answer is *yes*. According to Hoyningen-Huene: "The difference between the revolutionary discovery of new phenomena or entities and discovery that occurs in normal science is that the latter case in general leads neither to a revision of prior explicit or implicit knowledge, nor to an alteration of conventional experimental techniques nor to any correction in ways of interpreting data".[17] Kuhn himself says: ". . . scientific revolutions are here taken to be those non-cumulative developmental episodes in which an older paradigm is replaced in whole or in part by an incompatible one".[18] One analogy between scientific and political revolutions is the sense common to both that existing structures are malfunctioning; this, Kuhn goes on to say, applies also to the small changes associated with the discovery of oxygen and X-rays.

There is nonetheless some tension here. On the one hand, in Section IX of *The Structure of Scientific Revolutions*, from which the above quotation is taken, Kuhn assimilates anomaly-driven discovery to a case of revolutionary change (despite noticing the strain of so doing). On the other hand, in Section VII he emphasizes the distinction between anomaly-driven discoveries and changes in paradigm theory. The former were "not responsible for such paradigm shifts as the Copernican, Newtonian, chemical, and Einsteinian revolutions. Nor were they responsible for the somewhat smaller, because more exclusively professional, changes in paradigm produced by the wave theory of light, the dynamical theory of heat, or Maxwell's electromagnetic theory".[19] It is worth remarking that in Section VI "Anomaly and the Emergence of Scientific Discoveries", the chapter in which Kuhn discusses anomaly-driven discoveries, revolutions are *never* mentioned, whereas paradigm shifts are. An alternative, revised version of Kuhn would regard not every paradigm change as a scientific revolution. In terms of paradigms there are small shifts and there are radical replacements. Revolutions are those revisions to a paradigm that involve the replacement of old theories by new ones; anomaly-driven discoveries are non-revolutionary revisions requiring merely modification of existing theory or practice.

In this view anomaly-driven discovery is not attended by crisis. It may well lead to surprise or even opposition (Kuhn reports that Lord Kelvin thought that X-rays were a hoax). That something goes "wrong with normal research"[20] and is believed to have gone wrong need not be enough to precipitate a *crisis* in any normal sense of the word. (Again it is worth noticing that Kuhn mentions crisis in connection with changes that result in the emergence of new theories but not in regard to anomaly-driven discovery.) There may be a small-scale revision that fixes the problem and which is immediately acceptable to most scientists. Although Kuhn does not say as much, it is plausible to suggest that finding the solution may not require techniques beyond those in regular use in normal science. An issue we shall have to discuss is whether elements of a paradigm can be created as a consequence of advances in normal science. There seems no reason why a well-entrenched result of normal science should not create the sort of expectations characteristic of a paradigm. And if so, an experimental outcome that is in conflict with such expectations would be anomalous. In which case we might imagine that a wrong turning in normal science might be correctable by the techniques of normal science. Roentgen's X-ray case might be like this.

This is in part a proposal for reserving the word "revolution" for those more significant innovations involving adoption of a new theory. Normal science would then include some cases of belief revision, where those were revisions that involved no crisis and could be made by employing the theoretical and other techniques already available in that normal science tradition.[21] This suggestion notwithstanding, the "standard" Kuhnian definition

$$\text{revolution} = \text{any change to paradigm}$$

prevails. To mark it I shall write "revolution$_K$", so that we will be able to use "revolution" for our normal concept of revolution in science. This is important because the discussion of anomaly-driven discovery reveals that in addition to big revolutions$_K$ that are full-scale replacements of theory – such as the adoption of Dalton's atomic hypothesis – there are very small revolutions$_K$ – such as the discovery of X-rays – that are very different. The adoption of a small revolution$_K$ may involve no crisis and may look to many rather more like the acceptance of an ordinary normal scientific puzzle-solution than an intellectual upheaval. Small revolutions$_K$ are not revolutions in our normal sense. Later we shall ask whether there are revolutions (in the normal sense) that are not revolutions$_K$.

Crisis

Crisis is more than just the existence of an anomaly. It requires a sufficient weight of accumulated anomalies that a sense of professional insecurity is generated. That insecurity consists of doubts about the existing core theory and its attendant procedures and techniques, and also, therefore, about the reliability of the achievements of normal science. Hitherto a failure to solve a puzzle reflected negatively on the scientific capacities of the scientists who had tackled it. During crisis the blame shifts from the scientists to the paradigm. Returning to the subject of his first book, Kuhn points out that by the time of Copernicus, Ptolemy's astronomy along with Aristotelian physics were in such a state of crisis. There had never been an exact fit between Ptolemy's system and the best observations. Attempts to improve the fit were the stuff of Ptolemaic normal science. But puzzle-solutions involving adjustments to Ptolemy's system of epicycles and equants just led to ever-increasing complexity, and did so to such a degree that those who understood the matter doubted whether such a system could really describe nature or have been devised by God.[22] Kuhn admits that other forces were at play that helped precipitate the crisis, such as the waxing of Neoplatonism and the need for calendar reform. But, he says, the crux of a crisis is technical failure. And, as has been noted, calendar reform promoted crisis by highlighting the technical failures, especially the problem of the precession of the equinoxes.

Kuhn urges that not every observation that seems to be in conflict with a core theory must be seen as a counter-instance to it. Most failures of exact fit between theory and nature have the status of the puzzles of normal science. Newton's program of celestial mechanics showed a large range of imperfect fits. He himself had shown how to improve some and his successors others. The remainder were puzzles for yet later researchers to tackle. So not even a sizeable number of "imperfections" warrants a crisis. A crisis is brought about by more disruptive, disconcerting imperfections. Kuhn cites four kinds of more serious anomaly: (i) an anomaly that directly challenges a key theoretical generalization; (ii) one that inhibits an application of special importance; (iii) one that emerges as a result of a development in normal science; (iv) one that resists repeated attempts at solution. (These need not exhaust the list of kinds of serious anomaly.)

Revolution and theoretical innovation

The seriousness of such anomalies demands attention. Some can be handled by small-scale revision of the core theory. Thus we might expect scientists to try small, local adjustments or ad hoc amendments to the core theory in order to retain something like that theory while also eliminating a particular failure of fit. The solutions may not be entirely satisfactory and appealing, although some may be and the crisis is resolved. Competing solutions may be devised. And if there are several distinct anomalies, local revisions each made with one or two anomalies in mind will lead to a plethora of revised versions of the theory, each competing with the others, none able to solve all the extant problems. Such a proliferation of theories is, says Kuhn, a symptom of a crisis,[23] as exemplified by the crisis in phlogiston theory which generated multiple versions of that theory, where attempts to accommodate evidence were accompanied by increasing vagueness and decreasing applicability. This circumstance was ripe for the development of a theory radically different from the existing core theory. The development of a new theory with the power to attract adherents away from the old one is the start of a scientific revolution.[24]

Not all crises, however, culminate in a revolution. Normal science may eventually provide the resources required to resolve a crisis in the same way as a minor anomaly. Or, at a different extreme, there is not even a very radical new theory forthcoming that scientists find a satisfactory replacement.[25] Neither of these result in theory change – the interesting cases are the revolutionary ones.

In the event that a radical new theory is proposed, how does it come to be accepted, replacing its predecessor? There can be formidable obstacles in the way of radical theory change. New theories encounter considerable resistance. Established scientists will be reluctant to reject the theory that governed the work that made their reputations. In discussing the reception of Einstein's theories of relativity, I. Bernard Cohen quotes adversaries such as Louis Trenchard More, dean of the graduate school of the University of Cincinnati, who wrote that the persistence of relativity theory "will cause the decadence of science as surely as the medieval scholasticism preceded the decadence of religion".[26] Kuhn himself quotes Max Planck: "a new scientific truth does not triumph by convincing its opponents and making them see the light, but rather because its opponents eventually die, and a new generation grows up that is familiar with it".[27] As Cohen remarks: "the profundity of a revolution in science can be gauged as much by the virulence of conservative attacks as by the radical changes in scientific thought it produces".[28]

Yet it should not be thought that resistance is just a matter of old conservatives defending their interests against young radicals. There are deeper reasons, to be found in the nature of science itself. First, resistance is bound up with the very essence of normal scientific research.[29] Kuhn repeatedly emphasizes that in normal science the core theory, like other aspects of the paradigm, is just not in question. If the theory were potentially falsifiable, if it were not taken as a given, the activities of normal science could not proceed. This is why normal science seeks neither to test nor to confirm its core theories. Consequently resistance to the rejection of the core theory is built into the practice of normal science. Normal science requires that we see a failure of fit as a puzzle rather than as a counter-example.

Secondly, as Kuhn also stresses, the superiority of the new theory is not susceptible of proof.[30] The conservative may point to past successes of the normal science tradition as evidence that it may overcome the anomalies that the radicals regard as reason to adopt the new theory. Since the latter is in a more or less nascent state with no extensive track record, the comparison of old theory and new is not the comparison of like with like. It is comparing past achievement with future promise.

The problems of comparing radically different theories are of special significance for Kuhn. These are the sources of incommensurability – the lack of a common standard of evaluation. First, antagonists in a scientific debate may disagree with regard to the problems that are relevant and over the standards to be applied to proposed solutions. Kuhn points out that there was resistance to Newton's theory of gravity on the grounds that action at a distance remained an unexplained mystery, while after Newton's triumph that sort of concern was ignored or dismissed as metaphysical. Similarly, one might add, Einstein's resistance to the development of probabilistic quantum mechanics on the grounds that God did not play dice with the universe, or, less metaphorically, nature could not be irreducibly indeterministic, may seem old-fashioned or wrong-headed to those brought up in a tradition where indeterminacy is treated as a basic fact of life.

Secondly, radical theory change is accompanied by conceptual shifts that prevent unproblematic comparison. It might seem that Einstein's theories of relativity are straightforward improvements on their Newtonian predecessor. In particular the latter might be thought of as a special instance of the former, covering cases where velocities are low compared to the speed of light.[31] Using this view one could say that the successes of Newtonianism can be seen equally as supporting the Einsteinian theories, since the two coincide very closely in those cases, while the Einsteinian theories show an advantage when experiments are performed on cases where the theories diverge (high velocities, strong gravitational fields). The theories seem directly comparable, with Einstein's being better than or an extension of Newton's. But this, says Kuhn, all depends on the assumption that the key theoretical expressions have the same meanings and references in the two sets of theories. Kuhn denies that this is the case – the Newtonian conception of mass as a conserved quantity differs from the Einsteinian conception where mass is interchangeable with energy. "To make the transition to Einstein's universe, the whole conceptual web whose strands are space, time, matter, force and so on, had to be shifted and laid down again on nature whole."[32]

The third, and according to Kuhn, "most fundamental aspect of the incommensurability of competing paradigms", is the fact that, "In a sense that I am unable to explicate further, the proponents of competing paradigms practice their trades in different worlds". In a later chapter we will consider whether any useful philosophical sense can be made of this idea. For the moment, however, we may reflect that there is an everyday sense of "world" that chimes with what Kuhn is saying. We may have occasion to say of someone who has suffered a deep loss that "their world has fallen apart". And this is meant largely as a *psychological* reflection. In a similar vein we may talk of the world of a *kind* of person – the child's world, the world of a poet, the world of the professional soldier. In so doing we are not talking simply of the typical environment of such people. Rather we are referring to a characteristic outlook they may have. And we may describe this outlook by saying what the world contains; so, still in a colloquial way of talking, we may speak of the world of the medieval peasant as being inhabited by witches, ghosts, spirits and angels. Similarly, Kuhn says that the world of the Aristotelian physicist is one in which a swinging object is one which is falling, albeit in a slow and constrained manner, while no such thing is in the world of Galileo, where there is instead a pendulum whose motion is repeated again and again. This colloquial sense of world can be enough to express the thought that people whose worlds – their outlooks – are very different will find difficulty in communication and mutual understanding. And if the worlds of scientists may differ, then this incomprehension will be an obstacle to comparing theories originating in these different worlds.

"Incommensurability" does not *mean* "non-comparability"; nor does Kuhn think that incommensurability straightforwardly *entails* non-comparability. Proponents of new theories can make claims on their behalf that involve or invite comparison with predecessor or competitor theories. They can point to problems with the old theory that are solved by the new one. They can point to the successful prediction by the new theory of new phenomena, as well as an impressive quantitative fit between observation and prediction. Kuhn is not arguing that such facts give no reason for preferring the new theory – his claim is that they provide no *compelling* reason to adopt the new theory, they can never amount to *proof*.

Kuhn's view seems then to be this. There can be scientifically rational grounds for theory comparison and preference, as there are for comparison and preference among puzzle-solutions in normal science. In particular, one paradigm candidate may solve more puzzles than another and may be especially successful in dissolving the anomalies that caused the crisis while conserving many of the achievements of the previous paradigm. Compared to a simple quasi-Kuhnian model, where revolutions involve the complete overthrow of a paradigm and its replacement by something entirely different, this more sophisticated view allows a much greater degree of similarity between normal and revolutionary science. Old paradigms do play a part in determining the nature of their successors; the new paradigm must have some similarity to its predecessor, just as a normal puzzle-solution must. So what is characteristic of revolutionary research is that the evidence cannot be sufficient to compel rational assent to the new hypothesis – it is always open to the acolyte of the older theory to argue that the older theory will in due course solve all the puzzles that the innovating theorist regards as anomalous counterexamples. No new theory is perfect, and so the traditionalist will often be able to point to problems in the new theory that are absent in the old.

Since the move to a new theory is not a matter of rational compulsion, Kuhn likens it to a religious conversion. Deciding to support the new and largely untried theory will be a leap of faith. What then might induce a scientist to make the leap? In *The Structure of Scientific Revolutions*, Kuhn says that this is a little studied area worthy of further research. But he does list factors of the kind that are relevant. Some may be entirely particular to the individual, i.e. his or her idiosyncrasies, personality, or psychology. Some people are just more conservative or risk-averse than others. Recent research suggests that first-born children, even when adult, are less likely to be receptive to new or radical ideas than later born children. In the reception of Darwin's theory of evolution through natural selection, enthusiasm for the new ideas is significantly well correlated with being a younger child.[33]

Sometimes the individual will have other commitments that dispose them one way or the other. Kuhn refers to research that links Kepler's adoption of Copernicanism to Sun-worship. At the same time there will have been many who rejected Copernicanism because of a perceived clash with religious doctrines. Luther was quick to dismiss heliocentrism on such grounds. Other commitments that may exert an influence, consciously or otherwise, include nationalistic sentiment or respect for the reputation of the innovator. Kuhn lays especial emphasis on what may loosely be regarded as aesthetic grounds. We may link this directly to the quasi-religious nature of conversion. Just as Max Weber refers to charisma in connection with a decision to follow a religious leader, a new idea may attract adherents through its inarticulable aesthetic properties. A theory may just feel right; its simplicity and elegance may demand credence from some scientists in a way that cannot be expressed and which goes beyond the level of support offered by the evidence. Among the factors that influence the preferences of scientists we can distinguish between those factors that are internal to science – such as the nature of the evidence, the aesthetic appeal of the theory, its ability to solve problems, its affinity to existing puzzle-solutions – and factors that are external to science, which would include national sentiment, personal or religious bias, political and economic interest and so on. As we have just seen, Kuhn does mention external factors, and it is these that have been of greatest interest to post-Kuhnian sociologists and (some) historians of science. But for Kuhn himself it is clear that the strongest sources of influence on a scientist's thinking come from within the practice of science.

Critical concerns with Kuhn's picture

So far in this chapter I have articulated the pattern of the historical development of a science as Kuhn perceives it. This exercise has therefore been largely descriptive. Kuhn also has a theory based on the concept of a paradigm that seeks to explain the pattern thus described. This I shall spell out in the next chapter. But before we reach that point we should ask whether Kuhn's description is accurate. If it were badly mistaken the explanatory theory would necessarily be redundant, indeed mistaken. If there is no phenomenon, no theory is needed to explain it, and any theory that tries to do so will be erroneous – which is why there is no good theory of cold fusion or of inherited acquired adaptation. In my introductory sketch of Kuhn's system I drew a simple picture with the following components: immature science leads to mature science, which has two phases, normal science and revolutionary science; the corresponding explanation looked similarly simple and attractive. The move from immature science to mature science is the adoption of a first paradigm; normal science is science governed by a paradigm; revolutionary science is science without a governing paradigm – when the old has broken down but has yet to be replaced. But as we have already seen in this chapter the truth is rather more complex than this simple story suggests. The consequent concern is that the paradigm-based explanation will either be inadequate to the facts or too complicated to be especially illuminating. I shall look at three concerns of this kind, in increasing order of severity.

The first concern is that the normal science–revolutionary science dichotomy does not do justice to the variety that exists among changes in scientific thought, even if we exclude immature science from consideration. More specifically there may be no important distinction between normal and revolutionary science; there may instead be a continuum from small, insignificant cumulative additions to belief through moderately important changes involving a fair amount of belief revision to epoch-making revolutions. If true that would not of itself remove the possibility of an explanation in terms of paradigms, but the paradigm–no paradigm model for normal–revolutionary science would not do.

The second concern is an extension of the first. Let us assume that scientific changes are distributed over a continuum of greater or lesser significance. It is natural then to want to know the scale and shape of this distribution. Are the extremes very far apart, as Kuhn would have it, or close together? Do most changes fall into one of two groups clustered towards the extremes (a bimodal distribution – U-shaped or two-humped)? Or is there a single hump, near the middle? Or are most changes near one extreme, tailing off towards the other like a ski-jump slope? These statistical-sounding questions are heuristics for depicting different models of scientific change. The simple Kuhnian model would give rise to a bimodal distribution with a fairly broad base, i.e. two basic kinds of change that are clearly distinct. If, on the other hand, one's study of history suggested a narrow unimodal distribution then one would want an explanatory theory that posited one kind of change without too much latitude for difference between the least and greatest changes.

The third and most serious concern is that Kuhn's picture has no place for some classes of scientific discovery. So far I have gone along with Kuhn in equating the significance of a change with its being revisionary. Again on the simple model normal-scientific changes are small and cumulative while revolutionary changes are big and require revision or rejection of existing belief. Can there not be small changes that are revisionary (in a small way)? More importantly, why can there not be big changes that are not revisionary?

A continuum of changes?

I shall now look at these concerns in more detail – is there evidence to support them? The first suggested that there might be a continuum of changes from insignificant to revolutionary with instances of the various intervening degrees of magnitude. Using examples drawn mainly from Kuhn we can construct the following list of types of change or response to problems:

   (i) simple cumulative, no revision required;
  (ii) discovery driven by minor anomaly, requiring revision to non-paradigm beliefs;
 (iii) discovery driven by minor anomaly, requiring revision to paradigm beliefs;
  (iv) minor anomaly, not solved and shelved;
   (v) serious anomaly, solved within paradigm (i.e. normal science solution);
  (vi) serious anomaly, not solved and shelved;
 (vii) serious anomaly, solved with minor revision to paradigm;
(viii) serious anomaly, solved with major revision to paradigm.

The distinction between a serious and a minor anomaly is that between an anomaly that has a crisis-provoking capacity and one that does not. On a caricature of Kuhn's position it might be thought that all anomalies are potentially crisis-provoking – if there are enough of them at least. But as we have seen, Kuhn thinks that there are qualitative differences between "everyday" and crisis-provoking anomalies: "if an anomaly is to evoke crisis, it must usually be more than just an anomaly".[34]

Let us look first then at the responses to minor anomaly. Case (i) is just standard puzzle-solving. Case (ii) arises because not every scientific belief is a constituent of a paradigm. A puzzle may seem to be solved in the course of normal science, but then we discover a problem with it and must replace it with another solution. Case (iv) is also uncontentious. A puzzle may be set aside for later consideration. Indeed this may be done even in the case of serious anomalies.[35] What then about case (iii)? Can we talk of a minor change to a paradigm – aren't all changes to a paradigm major ones, being revolutionary?

Kuhn's standard definition says that a revolution$_K$ is any revision to a paradigm. A question that is in background of the discussion for the remainder of this chapter is whether this is a satisfactory definition. Does it match our ordinary usage of the term "revolution"?[36] We should not be misled by the apparent gravity of the word "revolution$_K$" that is implicit in "revolution". It is possible for there to be revolutions$_K$, i.e. revisions, that are small and involve no crisis – Kuhn's example of Roentgen's discovery of X-rays being one such. As we have seen, it involved a non-theoretical revision to the paradigm. It did provoke resistance but not a crisis, and the revision was quickly assimilated without a struggle between competing paradigms. Might there not also be such revisions that do not meet with resistance, that are uncontested, being immediately acceptable to all scientists in the field? Hubble's discovery of the expanding universe is interesting in this respect. It was more extensively revisionary than the discovery of X-rays, upsetting deep-seated beliefs about the structure of the universe. At the same time, although the discovery provoked a crisis in the sense of creating doubt about important elements of the existing paradigm, it also resolved that crisis, because Hubble's data and conclusion were both convincing. His proposal met no prolonged or serious resistance.

Furthermore, I suggest that Kuhn himself admits such cases in his discussion of discoveries. The latter are discussed in Section VI of *The Structure of Scientific Revolutions* ("Anomaly and the Emergence of

Scientific Discovery"), while crisis-evoking anomalies are discussed in the subsequent two sections (VII "Crisis and the Emergence of Scientific Theories" and VIII "The Response to Crisis"). In the opening paragraph of Section VII Kuhn says, "Discoveries are not, however, the only sources of these destructive-constructive paradigm changes. In this section we shall begin to consider the similar, but usually far larger, shifts that result from the invention of new theories".[37] The implication of this is that the discoveries discussed in Section VI include minor paradigm changes, smaller than the theoretical changes that (may) result from crisis.

Let us now turn to the serious, which is to say crisis-provoking, anomalies. Kuhn lists three responses to such anomalies, which were discussed above on p. 44.[38] These were (a) normal science resolution – which is (v) in my list; (b) no solution, problem shelved – (vi) in my list; (c) emergence of a new candidate for paradigm – which is closest to (viii) in my list. This leaves case (vii), a minor change to the paradigm. If we accept a minor change to the paradigm as a possible response to a minor anomaly, why not also in response to a major anomaly? Note that Kuhn allows a normal science response to a major anomaly, (v), as well as a full-blown revolutionary replacement of paradigms, (viii). So we might expect some resolutions of a crisis to fall somewhere in between.

Kuhn gives no argument, historical or theoretical, to the effect that these cases (iii) and (vii) do not exist. It is tempting to see Kuhn's picture of change as delineating a simple and powerful model. There are two sorts of change: small, cumulative ones during normal science, where no revision to paradigm beliefs occurs at all; and there are revolutionary ones, which because they occur as the result of a build-up of anomalies are large-scale replacements of theory. This is attractive not just because of its clean simplicity but because it is analogous to many phenomena where small changes under conditions of resistance lead to explosively large changes: the mathematical theory of catastrophe theory was designed to model many such phenomena, from mechanics (the straw that broke the camel's back), through the psychology of animals under stress, to developmental biology.[39] This polarization is also suggested by the sudden switching implicit in Kuhn's use of the gestalt analogy (gestalt shifts are also modelled by catastrophe theory).

My argument suggests that intermediate changes, paradigm revisions that are not fully revolutionary, in response to serious as well as minor anomalies, ought to be possible. Kuhn himself mentions one: "during the sixty years after Newton's original computation, the predicted motion of the moon's perigee remained only half of that observed. As Europe's best mathematical physicists continued to wrestle unsuccessfully with the well-known discrepancy, there were occasional proposals for a modification of Newton's inverse square law".[40] These then would be proposals for minor paradigm revision. But Kuhn plays down this intermediate case, adding that "no one took these proposals very seriously, and in practice this patience with a major anomaly proved justified. Clairaut in 1750 was able to show that only the mathematics of the application had been wrong and that Newtonian theory could stand as before". Kuhn underestimates both the level of concern and the significance of the proposals for change. The greatest mathematician of the era, Leonhard Euler, declared: "It appears to be established, by indisputable evidence, that the secular inequality of the moon's motion cannot be produced by the forces of gravitation".[41] The revisionary proposals came not from minor players but from Clairaut and Euler themselves. Imagine that Clairaut had not succeeded in resolving the problem, and that Euler's opinion that Newton's equation for gravitation could not explain the moon's motion had become widespread. The anomaly, which Kuhn admits is "major", would be clearly of the serious, crisis-evoking kind. Now imagine that one of the mathematical revisions gave a perfect fit with the observations, and furthermore that other anomalies (such as the anomalous motions of Saturn and Jupiter) had been resolved by that revision. It is reasonable to suppose that the scientific world would have come to accept the change to the central theory without prolonged resistance. Such a change would not be revolutionary in any major sense, since no changes in standards, concepts, or worldview are required. The exponent 2 of r in Newton's equation plays no special theoretical role (which it does in Einstein's version).

It seems then that we have reason to admit moderate changes that revise paradigms but not in a dramatic way that involves extended crisis, resistance, incommensurability and so on. This raises the question: where does the boundary fall between normal and revolutionary science? My typology of change involved two parameters – cause of change (non-anomalous puzzle, minor anomaly, serious anomaly), and result of change (cumulative addition, shelving, non-paradigm revision, minor paradigm revision, major paradigm revision). A definition of "normal" and "revolutionary" in science could make use of either of these parameters. On the simple model, non-anomalous puzzles and minor anomalies lead to cumulative additions and non-paradigm revisions, while major anomalies lead to major paradigm revisions. In which case a definition of normal science in terms of causes (non-anomalous puzzles, minor anomalies) would correspond to one in terms of effects (cumulative additions and non-paradigm revisions). But as we have seen, things are not so simple, for example a minor anomaly can lead to a (minor) paradigm revision. Hence we have to choose between the two parameters in framing our definition of normal science, or employ some considered combination of them.

The standard definition takes the effect to be the criterion – normal scientific change is change that requires no revision at all to the paradigm. The standard definition must face up to the fact urged above, that there are intermediate levels of more or less moderate revision to core theories. To call all such changes "revolutionary" is to part company with our everyday notion of a scientific revolution. Could the causal criterion do the job alone? Then normal science would be any episode that responds to a non-anomalous puzzle or minor anomaly. By that token any episode resulting from a major anomaly would be revolutionary science. The problem with this is that major anomalies can sometimes be resolved by normal scientific techniques.[42] It would also be odd to call such events revolutionary. This is one reason why I made the suggestion in Note 21 that we restrict the term "revolutionary" just to those instances of major theory revision, and use "extraordinary" for cases of minor revision such as anomaly-driven discoveries like Hubble's and Roentgen's. But it seems that whatever definition we employ it remains the case that the normal science versus revolutionary science dichotomy cannot do justice to the variety of episodes in science. Kuhn's terminology gives an artificial sense of there being two quite distinct kinds of scientific change. Reflection on this variety suggests that the distribution of episodes is not bimodal but instead shows a greater degree of continuity, with intermediate cases being not especially less frequent than the extremes.

How different are normal and revolutionary science?

I shall now turn to the second, very much related concern, that the differences between revolutionary and normal science are not as great as Kuhn depicts. First, even if it is accepted that the most extreme cases, such as mundane laboratory work (blood testing) versus deep and grand innovations (the theory of quanta) are very different indeed, the fact (if it is accepted as such) that there is a continuum of cases between those would indicate (a) that the "average" piece of normal science and the "average" revolutionary episode are not nearly so far apart, and (b) that the most innovative normal science will not be readily distinguishable from the least radical revolutionary science.[43]

One feature of revolutions (or revolutions$_K$), as Kuhn describes them, that makes them seem very different from other moments in science is that they involve incommensurability. If the incommensurability of theories of paradigms meant (a) that they cannot be compared, and (b) that proponents are mutually incomprehensible, then that would certainly make revolutionary change vastly different from the normality of science where we think that competing solutions to puzzles are comparable and their supporters able to understand one another. Talk of "world changes" also helps make revolutionary shifts seem highly esoteric. Incommensurability and world changes are topics for later chapters. Nonetheless it has already been demonstrated that Kuhn means nothing so dramatic by incommensurability in *The Structure of Scientific Revolutions*, and we shall later see that he came to endorse an even weaker notion. Incommensurable theories (or paradigms) may well be comparable. The comparison is just not rationally decisive. Is it the case then that all normal scientific changes are rationally decisive and all revolutionary ones are not? Is there a clear dichotomy between the rationally decisive and that which is not? Or is it better to conceive of arguments for change as being rationally compelling in different degrees?

Kuhn in fact says nothing about puzzle-solving that suggests that accepted solutions must be decisive. In Kuhn's view, the acceptability of a proposed puzzle-solution is a matter of its resembling an exemplary puzzle-solution. This is something we might well expect to come in degrees, and the story has little to say about counts as *rational* or *decisive* in a puzzle-solution. Indeed, there is no particular reason why a solution should be decisive to become accepted. Perhaps it is just the best available solution and none other has been suggested. In terms of Kuhn's analogy, we might accept a crossword puzzle answer without finding it rationally decisive. At the same time, not every revolutionary change is open to prolonged rational dispute. One cannot rationally cling to a flat Earth hypothesis, pre-quantum classical theories of the atom, phlogiston theory, creation science and so on, despite knowing all the evidence of modern science. Furthermore, given the standard definition of "revolution$_K$", we must include quite minor theoretical and even experimental changes as revolutionary$_K$. Were arguments for the existence of "X-rays" rationally undecidable while the results of a novel piece of normal science, such as Cavendish's determination of G, unarguable? Did rationality require acceptance of Coulomb's law but not of Hubble's? It looks as if the difference as regards incommensurability of standards is one of degree not of principle, and that innovative normal science is not in this respect different from science that results in moderate changes to core theory or to central techniques.[44]

Similarly, conceptual or semantic incommensurability does not mean anything so dramatic as mutual incomprehension. If it did, the work of an historian of science like Kuhn would be impossible. The issue is one, rather, of the translatability of concepts. I shall argue later that translation failure is no big deal. Most words and expressions we use, scientific or not, cannot readily be translated into other terms of English. Hence almost any introduction of a new concept will involve failure of translation into the previously existing vocabulary. So any distinction between normal and revolutionary science based on this criterion will not demonstrate that they are really quite different in kind. In any case, even if much normal science involves no conceptual innovation, it cannot be true that none does. Names for newly discovered stars and nebulae will be trivial cases. But more sophisticated innovation may occur too, as a result, for example, of the discovery of new empirical laws. Coulomb's law permits the introduction of the concept of "electric force" (and hence of "electric potential" and "electrical field" and so on).

Kuhn would distinguish between such cases of conceptual addition and those involving more radical conceptual rearrangement, the latter being characteristic of revolutions. For example, he thinks that the Einsteinian revolution involved the wholesale reorganization of our conceptual economy. That would indeed be dramatic, but it is by no means obvious that anything of the sort actually occurred. Even if it is true that in the Einsteinian case there was radical conceptual change, such events would in any case be rarities. The same would be true of theoretical changes involving a change of "worlds". The gulf between revolutionary and normal science is exaggerated if we concentrate on events that have occurred only once or twice in the history of a science and affect the whole or large portions of the science. Thus in terms of root and branch conceptual revision and change of world-view we can in physics identify the scientific revolution of the sixteenth and seventeenth centuries initiated by Galileo and completed by Newton, and the relativistic and quantum revolutions in the first decades of the twentieth century. But the definition of "revolution" we are given is far broader than this. It is not at all clear that Maxwell's towering intellectual achievements required a corresponding conceptual revision or rearrangement (as opposed to adding to existing concepts). The pedant might try to argue that Maxwell's concept of "field" differed from Faraday's, or that the theory of light as electromagnetic waves required revision to the concept of "light". This is not a debate that we can pursue here in any detail, but two brief comments might point us in the right direction. First, as regard the "field" concept, even if there was a change it was not a change of any significant kind that might have introduced deep problems of incommensurability. Secondly, as regards the concept of "light" we need to guard against an equivoca-tion between changes in concept (i.e. meaning) and changes in conception (i.e. theoretical belief). Not every theoretical innovation and not even every theoretical revision requires change to existing concepts (and in particular not to the networks of concepts). We can still mean the same by the word "light" even if we have changed what we think light *is*.[45]

In chemistry the revolutions of Lavoisier and Dalton did involve conceptual revision, but subsequent revolutions have not required anything like the same degree of conceptual change. The two most significant theoretical innovations in chemistry were Mendeleev's periodic table and the quantum theory of the atomic bond. While certainly revolutionary in that they opened up new areas of research, neither advance required a "whole conceptual web . . . to be shifted and laid down again on nature whole".[46] Similar remarks may be made about the discovery of the structure of DNA and the theory of plate tectonics, examples to which we shall return shortly. Even further from radical conceptual change were the cases of Roentgen's X-rays, Lavoisier's oxygen, or Clairaut and Euler's proposals for amendments to Newton's law of gravity. *At most* only a small number of scientific revolutions or revolutions$_K$ involved incommensurability of a significant kind. Consequently the majority of revolutions$_K$ do not differ from normal science in this respect.

Conservative revolutions and revolutions without crisis

Since a revolution$_K$ is just any revision to a core theory or practice, necessarily conservative, non-revisionary changes belong to normal science. Furthermore Kuhn prefers to think of revolutions$_K$ as being presaged by crises, although as we have discussed Kuhn ought and perhaps does accept the existence of minor paradigm shifts that are attended at most by resistance of less than crisis proportions.

Even if we think of the more obvious revolutions, not every one follows a crisis. Einstein's theory of general relativity was a revolution by any standard. But unlike the special theory it did not succeed a crisis that prompted it. Einstein himself claimed that if he had not discovered special relativity someone else would have done so (suggesting Paul Langevin as the likely candidate). But regarding general relativity, Einstein thought that it might well not have been generated by another mind. There was no anomaly whose solution prompted Einstein, no lacuna for the theory to fill. The only explanation for the theory was Einstein's own genius, seeing connections none had been seen before (e.g. between accelerated motion and the experience of being in a gravitational field). Although some physicists could not see the point of the general theory, many others did. The general theory was published in 1916, and the First World War must have delayed its acceptance. But that was not sufficient to prevent it being championed by the Plumian Professor at Cambridge, Sir Arthur Eddington, who gave the theory its famous vindication in the year following the armistice. Despite the profound conceptual shift Kuhn sees in the revolution of general relativity, it was a revolution (and a revolution$_K$) that was precipitated by no crisis and attended by little serious resistance.[47]

We might expect that if a signal scientific advance is brought about by a crisis consequent upon the accumulation of anomalies, that advance will be revolutionary$_K$, in virtue of being revisionary. But if there can be revolutions$_K$ without crises, might not that suggest that there could be great advances that we might ordinarily classify as revolutions that do not involve revisions, that are conservative with respect to earlier achievements, adding but not subtracting from the existing paradigm?

If such cases existed Kuhn's model, however elaborated, would fall apart. For then there would be revolutionary episodes that were not preceded or accompanied by the demise of a paradigm; some revolutionary science would be just as cumulative as normal science. It is not surprising therefore to find that Kuhn emphatically denied the existence of non-revisionary revolutions: "After the pre-paradigm period the assimilation of all new theories and almost all new sorts of phenomena has in fact demanded the destruction of a prior paradigm and a consequent conflict between competing schools of scientific thought. Cumulative acquisition of unanticipated novelties proves to be an almost non-existent exception to the rule of scientific develop-ment".[48] At first sight this looks highly implausible. It seems to suggest that at any given time all current paradigms taken together constitute a *complete* picture of the world – one that not only says what phenomena there are in the world but also leaves no room for any other kinds of phenomena. Thus any actual phenomenon is either encompassed by some paradigm or must come into conflict with it – there is no room for neutrality. "Paradigms" says Kuhn, "provide all phenomena except anomalies with a theory-determined place in the scientist's field of vision."[49] Everything is either expected or anomalous. For if paradigms were not complete, then it might be possible for us to discover something of great significance but which conflicted with no existing belief and so whose assimilation required no revision.

But why should paradigms be complete in this way? Could not a paradigm seek to account for a range of phenomena while being agnostic about the existence of other kinds of phenomenon? One would have thought that even a superficial understanding of the history of their subject would encourage most scientists to think that there is more under the Sun than they have dreamt of, so they had better not exclude that possibility in their theorizing.

It is perhaps possible to reconstruct a case for the non-existence of non-revisionary revolutions without depending on or requiring the completeness of paradigms. Kuhn can admit the possibility of phenomena that are neither predicted nor excluded by our para-digms as long as we do not get to perceive them. Kuhn's accounts of normal science and of perception suggest that this could be so. Obser-vation and experiment are not carried out serendipitously but with precise ends in mind. Scientists often already know the result to be produced, or will have a range of possible answers in mind. Hence the outcomes that they will notice are those that either accord with expectation or directly conflict with it. Scientists are not looking for facts not germane to the aim of the enquiry, so those will not come to their attention. This may be true even if facts worthy of investigation came into scientists' field of vision. Let us recall Fleck's example of agar plate with a multitude of bacterial colonies. There were many ways of grouping the colonies, but scientists will choose just one or two. So if scientists' aims are to record the proportion of the total that are colonies of bacterium X and if they know to associate X with smaller, light-coloured colonies, then they will quickly be able to identify the proportion as being 2 per cent and the experiment will be concluded. The fact that there is variation in size, transparency, colour and other properties among the remaining 100 colonies is of no interest to them unless they have some prior expectation about variation that is violated, in which case that fact might spring to their attention. But facts not under direct investigation nor counter to expectation will pass unnoticed, even though they may be the key to important potential discoveries. Unnoticed facts will not lead to revolutions but anomalous facts, which thereby make themselves seen, may do.

Even this case, I think, is overstated. Things certainly get noticed by being unexpected. But the unexpected doesn't have to be something we positively expected to be otherwise. And even where we do have expectations whose violation leads to significant facts being noticed, those expectations need not have been generated by a paradigm. We have non-paradigm beliefs including non-paradigm theoretical beliefs. And so if these conflict with an observation, the revolution that results, if any, may require no revision to any *paradigm*.

Most importantly, the range of possibilities considered as potential outcomes to a (normal) scientific investigation might not be as constrained or as specific as Kuhn suggests. The case of the discovery of the structure of DNA illustrates this. The problem was well defined, but the range of possible answers was at the outset only loosely constrained. For a long time it was not clear whether DNA contained two or three helixes – or was helical at all. But the discovery of the double-helix and the existence of base pairings, unexpected but not contra-expectation, were clearly revolutionary in their consequences for biochemistry and molecular genetics. A discovery that many regard as the most important of the century simply does not fit Kuhn's description of scientific development – it originated in no crisis and required little or no revision of existing paradigms even though it brought into existence major new fields of research.[50]

I shall conclude this section with some remarks about the nature of change in sub-disciplines. The impression that revolutions are major revisions brought on by crisis and so are very unlike normal science is fostered by focusing on a few revolutions such as those of Einstein, Dalton, Darwin, Newton and Lavoisier, that change an entire science and dominate it for many decades or centuries. Yet there is no reason to think that such a focus is justified by our pre-theoretical conception of a revolution, nor is it endorsed by the detail of Kuhn's discussion. Kekulé's uncovering of the structure of benzene transformed organic chemistry but not inorganic or physical chemistry. And within organic chemistry, the hypothesis of chiral (dissymmetrical) molecules created the field of stereochemistry without directly affecting other aspects of organic chemistry. The fact that fields, branches of fields and sub-branches have proliferated over time (a fact the recognition of which requires only a review of the increasing number of ever more specialized journals), implies that at points in history innovations and discoveries have occurred that have become the seeds of new disciplines and sub-disciplines.

The existence of mini-revolutions at the sub-disciplinary level is not itself inconsistent with Kuhn's normal/revolutionary science picture, since we can imagine that distinction operating at different levels. It would, however, require more detailed elaboration than Kuhn has given us.[51] How does a revolution at the sub-disciplinary level relate to normal science at the broader disciplinary level? More importantly, how does a science proliferate its sub-disciplines? We need to see how this can be reconciled with the essentially cyclical nature of Kuhn's picture. Furthermore, acknowledgment of the different levels of scientific change will need a corresponding accommodation in the explanatory theory of paradigms. The notion of paradigm will have to be able to make room for the idea of sub-paradigms, paradigms for specialities within paradigms for general disciplines.

Conclusion

What I have called the simple model divides instances of scientific change into two clearly distinct kinds: normal and revolutionary. The standard definition of "revolution$_K$" says that revolutions$_K$ are precisely those episodes that involve change to core beliefs, techniques and practices. Kuhn also thinks that revolutions$_K$ thus defined are usually also those changes brought about by a crisis and which engender incommensurability with earlier theories.

Such a model invites explanation in terms of paradigms, consensual, even dogmatic structures that ensure conservative change until great stress, in the form of crisis fomented by accumulation of anomalies, brings about their change. This explanation we shall investigate in the next chapter. But the explanation looks already that much weaker to the extent that the simple model is inaccurate.

While thinking in terms of the simple model is encouraged if we think especially of famous major revolutions, Kuhn does do greater justice to the intricacies of the historical facts and so adds much detail to the model. But at the same time this robs the model of much of its force. Instead of a clear dichotomy between normal and revolutionary$_K$ science there is a range of different historical episodes within each such that the most innovative normal science looks just as significant as the least radical moment of revolutionary$_K$ science. There is reason to think, on historical grounds, that all points on a continuum from mundane conservative accumulation to thoroughgoing revolution are well represented by the facts. Furthermore there is no great distance between a typical case of what Kuhn takes to be normal science and a typical case of revolutionary$_K$ science. The difference is a matter of degree not of kind. Next, crisis is not a necessary part of either revolutions$_K$ or revolutions, nor is crisis sufficient for them; and lastly, on any sensible conception of revolution, not all revolutions need be revisionary. Some revolutions can be conservative accumulations of new but significant discoveries and theories. Such revolutions are not explicable in terms of one paradigm failing, generating a crisis and then being replaced by a competitor. These revolutions may just be the blossoming of normal science.

The last point suggests that neither Kuhn's definition of scientific revolution$_K$ (as a revision to a paradigm theory) nor his description of the causes of a revolution$_K$ (as brought about by a crisis) match our ordinary, pre-Kuhnian conception of a scientific revolution. What then are scientific revolutions as we normally understand the term "scientific revolution"? I suggest that we cannot decide whether an episode is genuinely revolutionary either by seeing whether it is revisionary or by considering its origins. Rather we recognize revolutions by their effects on subsequent history of science.[52] Revolutionary are those changes in scientific belief that lay the foundations for significant and novel research – where "change" may be understood as including both revision to existing belief and addition to it. As a definition this is vague, but then our use of the expression "revolution" when applied to science is not precise either. And this vagueness does some justice to the fact that the boundary between revolutionary and non-revolutionary science is not clearly perceptible but lies somewhere on a continuum of shades of scientific change. Furthermore, if we construe "significant" as being a tacitly relational expression, it will make sense to ask "significant for whom (for which group of scientists)?"[53] We will then be able to do justice to the claim that revolutions can occur at different levels of discipline and generality as we will be able to describe as revolutionary a development that transformed research in some particular field even if it has little effect outside that field. Conservative or only mildly revisionary revolutions in sub-disciplines are the obvious mechanisms for the proliferation of specialisms and new areas of research. The suggestion that revolutions be characterized in terms of their historical consequences is not entirely un-Kuhnian, for the idea of laying foundations for future research could be explicated in terms of the institution of a new paradigm. The significant divergence from Kuhn is in the claim that new paradigms in a mature science do not have to be erected only on the ruins of old ones.

Chapter 3
Paradigms

Thought-styles in action

In the period preceding the First World War there were two theories of the aerofoil: the discontinuity theory and the circulation theory. The discontinuity theory favoured by British scientists was based on Lord Rayleigh's calculations of the force on a flat plate at an angle to a stream of fluid. The theory claimed that an aerofoil created an area of turbulence on the upper, trailing edge of the wing that would be at a lower pressure than the smoothly flowing air on the underside of the wing. The circulation theory, prevalent in Germany and Russia, claimed that the air molecules in the region of the aerofoil have a tendency to circulate around the wing. On the upper side this tendency reinforces the flow of the air while on the lower side it impedes it, creating a difference in pressure and hence providing lift.[1]

The circulation theory is the one universally accepted today – while the discontinuity theory is alarmingly close to current accounts not of flying but of stalling. This story of two theories in competition, one of which superseded the other, becomes interesting when we ask why exactly were the British theorists resistant to the circulation theory? For Lord Rayleigh himself had published the formulae on the which the circulation theory is based; in 1878 he showed that a moving tennis ball spinning about a vertical axis will tend to veer to one side, for precisely the reasons that the circulation theory said that the aerofoil created lift. Rayleigh's account was even incorporated into Horace Lamb's *Hydrodynamics*, a book well known to all the British researchers in the field since almost all of them had been trained in the mathematical tripos at Cambridge University where Lamb's book was a key text. So the lack of British receptivity to the circulation theory seems even more extraordinary. The explanation provided by David Bloor points to the nature of the Cambridge tripos examination. The latter had for decades brought to the fore some of Britain's greatest mathematical physicists, including Maxwell, Stokes, William Thomson (Lord Kelvin), J. J. Thomson and Rayleigh himself. Success in this most rigorous of examinations required correspondingly focused drilling in the kinds of question likely to be asked. Bloor's hypothesis is that this training created a mindset that partitioned problems according to their place in the syllabus. The circulation theory was not typically presented in a mathematically rigorous form, its supporters being more frequently engineers of a cast rather different from the mathematical theorists of Cambridge. But even when it was presented in a formal manner the circulation theory did not resemble any problem that was recognizable to a Wrangler – a first-class graduate of the tripos – as belonging to the aeronautical theory of lift. And since the circulation theory was presented as a solution in that field, the Wrangler's mindset either did not notice or did not see the relevance of the connection with the same formula employed in a different kind of problem, the problem of lateral deviation. According to Bloor the creation by training of a tripos mindset is not all there is to be said on the matter, for that way of thinking was reinforced socially. Not only did the majority of the leading members of the Advisory Committee for Aeronautics have that training, but many of them had fellowships at Cambridge. Interaction among these men kept the tripos way of thinking alive, an instrument of their everyday research not just a relic of student days.

No set of methodological or inferential rules were employed to show that the circulation theory was worse (or for that matter better) than the discontinuity theory. It is the mindset that explains a difference in theory preference. We have seen this idea before, in Fleck's *Denkstil* (thought-style) and it is to be found elsewhere, for example in Pierre Duhem's notion of national styles of thinking. Kuhn's notion of a paradigm similarly aims to capture the idea that a socially reinforced, trained way of thinking may better explain the historical development of science than the Old Rationalist emphasis on rationally justifiable rules and methods.

Paradigms as explanations

When compared with Fleck and Duhem, there are two significant advances in Kuhn's use of this idea. First, he has a particular account of the history of science that he thinks the thought-style idea can explain. Secondly, he has some specific ideas about what thought-styles are. This is why Kuhn's views should be regarded as a fully fledged theory. The case is similar to those of scientists whose observations lead them to discover a regularity in nature and who then turn to theory in order to try to explain the regularity. So one might first, through observation and experiment, notice that the pressure of a fixed quantity of gas varies in inverse proportion to its volume; one might then seek to explain this law in terms of molecules that collide elastically with one another and with the walls of the container. Similarly Kuhn claims to have discovered the pattern of normal science-crisis-revolution. His theory explains this pattern by positing the existence and nature of paradigms.[2]

Kuhn himself does not clearly distinguish between the explanatory and the descriptive parts of his project, and so the term "paradigm" is to be found in his descriptions as well as his explanations of change. For that reason I have had to use it too; where I have used the term it has been in the context of paradigm change, which has usually been the same as change in the core theory or allied (e.g. instrumental) commitments. But paradigms are not identical with the core theories of normal science, although there is a connection. What Kuhn did mean exactly by "paradigm" is not easily discernible. By 1969 he himself came to think that in *The Structure of Scientific Revolutions* he had used the word with two different senses. The kernel, at least, of the concept(s) of paradigm is fairly clear. What explains the possibility of normal science is the existence of a certain consensus among the community of scientists. It is this consensus that breaks down during crisis and which is rebuilt in the wake of a revolution. Characterizing a paradigm as a certain sort of consensus invites, as Hoyningen-Huene says, two further questions: (a) what is the consensus a consensus *about*? and (b) what generates and maintains the consensus?[3]

As regards the first question, Kuhn, in 1969, gave two answers corresponding to the two senses of paradigm that he identifies: "One sense of 'paradigm' is global, embracing all the shared commitments of a scientific group; the other isolates a particularly important sort of commitment and is thus a subset of the first".[4] The first sense is, says Kuhn, a sociological one; it is a very broad notion, encompassing "the entire constellation of beliefs, values, techniques and so on shared by members of a given community".[5] This he names a "disciplinary matrix" – *disciplinary* because it is what is possessed by all members of a given discipline and *matrix* because it comprises distinct elements.

What are the elements of a disciplinary matrix? Kuhn does not claim to provide an exhaustive list, but the key components are:[6]

- symbolic generalizations (formalized statements, covering both laws and definitions);
- metaphysical beliefs;
- scientific values (such as the desirable features of a hypothesis – Kuhn lists accuracy, consistency, breadth of scope, simplicity, fruitfulness);
- heuristic models;
- exemplary concrete puzzle-solutions.

This last component, for which Kuhn uses the term "exemplar", constitutes the paradigm in the second, narrower of Kuhn's senses, "the most novel and least understood aspect of this book" he says in the Postscript to the second edition of *The Structure of Scientific Revolutions*.[7]

Paradigms as exemplars are "a set of recurrent and quasi-standard illustrations of various theories in their conceptual, observational and instrumental applications. These are the community's paradigms, revealed in its textbooks, lectures and laboratory exercises".[8] The operation of paradigms-as-exemplars is seen clearly in Bloor's account of British adherence to the discontinuity theory. It is the training of the tripos men with puzzles and problems resembling examination questions that created their mindsets. Furthermore, the link between such training in puzzle-solving and productive research was very close. Some examiners would even set their own research problems as examination questions. It is hardly surprising that if examination questions were seen by examiners in terms of research problems, then conversely the Wranglers would go on to see their own research problems in terms of the puzzles with which they had been trained.

So what do paradigms-as-exemplars do? Hoyningen-Huene identifies three functions for exemplars:

(i) *semantic*[9] When performing normal science, framing and solving a puzzle, we use a variety of concepts, what Kuhn calls a "lexicon".

Some of the members of the lexicon are *empirical* concepts, i.e. concepts that can be fairly directly applied in observational or experimental circumstances. At one extreme colour words can be applied in many cases just on the basis of perception while highly theoretical notions like "quark" cannot be. The former are empirical concepts. One function of exemplars is to give such concepts their meanings. Without a concrete exemplary application of a concept we cannot get fully to understand it. (Imagine trying to explain colour concepts to someone without giving examples of samples of the colour.)

(ii) *puzzle identification* Exemplars help to identify puzzles for future research. And exemplars may also tell the scientist which puzzles are especially worth working on.

(iii) *research assessment* Exemplars provide standards by which we may judge whether a proposed puzzle-solution is any good.

Having identified these functions we will need to ask two further questions: (a) Are these functions adequate to explain the cycle of the history of science? (b) What exactly are exemplars? Can they really fulfil the functions ascribed to them? In order to answer these questions, especially (b), I need first to mention a fourth function of exemplars that is not explicitly mentioned by Hoyningen-Huene:

(iv) *solution identification* Exemplars fill a role that fits between function (ii) and (iii); they enable scientists to see the world in such a way that solutions to those puzzles become apparent.

A glance back at the analogy between scientific puzzle-solving and the solving of other puzzles such as crosswords illuminates these functions and their source. Taking function (ii) first, the analogy is admittedly fairly weak; recognizing a crossword puzzle in a newspaper as a puzzle does not seem to require a great deal of effort. But even so, we might imagine someone who had never seen a crossword puzzle being completely mystified. One needs *some* prior experience of crosswords to be able to recognize one as such (rather than as a decoration or a diagram). Furthermore just seeing a crossword beforehand is not sufficient experience – one would need to see how the clues relate to the grid and to each other. This sort of experience is best gained by observing someone actually doing a crossword puzzle. One might even argue that it is this experience that gives one an understanding of the *meaning* of "crossword puzzle", in an analogy with the semantic function (i). Similar remarks may be made at the level of individual clues. In simple crosswords all the answers are synonyms of the clues, but in more sophisticated puzzles there are a variety of different kinds of clue – anagrams, cryptic clues and so on. A novice simply cannot spot the hints in a clue that might indicate what kind of clue it is (c.f. function (ii) again). Correspondingly, novices are unable to set about tackling the clues of a cryptic crossword – they will not even know what to look for in a possible solution (c.f. function (iv)). Even when told the solution novices will have no feeling for *why* it is a good solution (c.f. function (iii)). It is only certain sorts of experience that give people the capacity to recognize the kind of clue a clue is, to set about solving it, and to know a good solution when a possible answer comes to mind. That experience starts, as suggested, with the novice following an experienced solver, listening to explanations of how the hints are recognized and why the solutions are appropriate. The novice may then attempt some of the easier clues – a feeling for the clues, hints and solutions is reinforced by getting these ones right, checked against the authoritative answers in the next day's newspaper, as well as by trying to get a sense of how those answers relate to their clues. Eventually the crossword solver may become so fluent in puzzle-solving that solutions come almost without thinking. Experts may not immediately be able to say how they arrived at the solution and may be more confident that it is right than about precisely why it is right.

In all this the explanations and solutions given by the expert or provided by the newspapers are exemplars in Kuhn's sense. They help us to understand what "crossword puzzle", "cryptic clue" etc. mean; they help us to recognize kinds of puzzle; they enable us to see possible solutions as good ones (or bad ones); they show us how to go about getting a solution. The crossword analogy illustrates a central feature of Kuhn's exemplar concept. Exemplars work by giving us a certain *feel* for puzzles, an *intuition* for solutions, and *sense* of the rightness of answers. In learning to solve crossword puzzles people do not learn a set of neatly expressible rules. That is not to say that they do not learn through verbal instruction – clearly they do; nor that there are no rules or methods that could be written down – there are some such. Rather, much of what is learned cannot be usefully transcribed.

The fact that the same exemplars have both functions (iii) and (iv) explains why Kuhn rejected the Old Rationalist dichotomy between the context of discovery and the context of justification. They held that the reasons given for accepting a certain belief or change in theory are to be distinguished from the causes that enabled the researchers to make the discovery in question. Ludwik Fleck rejected a sharp dichotomy of this sort and so did Kuhn. Exemplary puzzle-solutions not only provide a standard whereby solutions to other puzzles may be judged (c.f. function (iii)) but also, via their use in training, explain how it is that students and researchers faced with a puzzle can see it in such a way that a solution becomes apparent. The experienced crossword solver sees a clue as presenting an anagram of a certain word; the physicist sees a problem as requiring the application of Newton's laws in a particular form. The same recognitional capacities used in judging problem solutions are used in coming to them.[10]

Connectionism and intuition

At the same time Kuhn is not saying that puzzle-solving is something mystical or inexplicable.[11] There are explanations of these recognitional capacities, but the explanations are not in terms of learned rules but instead fall within the domain of neurophysiology. It is with *rules* that Kuhn wants explicitly to contrast exemplars. Some puzzles are soluble by learned rules. For example, we may learn that the solutions to a quadratic equation of the form

$$ax^2+bx+c = 0$$

are $[-b + \sqrt{(b^2-4ac)}]/2a$ and $[-b - \sqrt{(b^2-4ac)}]/2a$.

(Even so it may not be immediately obvious to all that

$$3x(x/y - 2.1) = (1.0697 \times 10^{-2})y$$

is a quadratic equation (in x/y) that may be solved with this rule.)

Yet not every puzzle that is soluble is soluble by such rules, even in arithmetic, where there are infinitely many theorems that are provable but which have no rule to tell us how to find such a proof. A very different problem for which rules are not available is pattern recognition, for example the problem of recognizing people's faces. We might try to help someone recognize a person by giving an explicit description of their features. But it would be misleading to think that all face recognition is essentially like that (with "internalized" descriptions).

Similarly it would be a mistake to think that scientific puzzle recognition, understanding, solving and evaluation may be achieved by following rules. The rules under consideration include explicit definitions of concepts, abstractly stated laws and theories, and explicit methodological precepts. Although he does not say so, Kuhn clearly has positivist and other Old Rationalist conceptions in mind when he talks of rules in contrast to exemplars. Carnap's view of science as consisting of formalized or formalizable theories, which include correspondence rules that provide meanings for theoretical terms, and whose degree of confirmation can be measured using inductive logic, exemplifies a view of science as rule-governed.

The sort of distinction Kuhn is seeking to make is much more familiar to psychologists and philosophers today than it was in the 1960s. Much of the advance in our understanding of this area has cognitive science to thank, in particular co-operation between computer science and neuroscience. In the 1960s computers first become widely used by scientists and academics. The capacities of the computer seemed to provide a model of human intellectual activity. The particular capacities of a computer depend on the software it runs. And the software or program is just a list of instructions or rules. Hence it was reasonable to suppose that human thinking can be understood as rule-governed. Furthermore, specific programs that aim to carry out human tasks are designed by trying to extract from human behaviour the rules that are supposed to govern it. So, for example, a program for buying a selling shares will have instructions that correspond to the (supposedly) rational rules of share trading – e.g. sell an equity that shows a prolonged drop in price. Since these rules are taken to encode the knowledge that a human expert has, the programs are known as "expert systems". In the first phase of classical Artificial Intelligence (AI) it was assumed that humans and computers should both be understood as expert systems. But in the late 1970s and 1980s research on the brain suggested that it is unlikely that human thinking always operates like this. If we delete some lines from a computer program we might expect to find that the program fails to function at all; or if the lines are from some rarely used sub-routine then the program will function perfectly until this sub-routine is called upon. In this way the functioning of a computer program is all or nothing. But human cognitive capacities are not like this. People who suffer minor brain damage find that their capacities show *partial* deterioration; they may find, for example, that they can do the same things but only more slowly, or that their performance is less accurate. In the jargon, brains "degrade gracefully". Only in severe cases of damage is a function lost completely. In less severe cases we also find that individuals may eventually recover full function, using some other part of the brain that previously had no role in this activity.

As a picture of what is going on, much better than the expert system analogy is the *connectionist* or *neural net* model. In this model there is no software or encoded expert knowledge; there are no rules. Instead there are many nodes, each of which connects to several other nodes forming a large, complicated network. These connections allow activity at one node to influence activity at connected nodes. Furthermore, these connections can be added or removed, and the strength of their influence can be increased or decreased. According to the connectionist model learning often takes the form of the evolution of such a network. In a given state of the network it may yield a certain output for some particular input. So if the input is a retinal stimulation from a painting, the output might be the name of a painter. Should the utterance be mistaken, then the network operator (which itself may be part of the overall system, as in the case of the brain) will randomly adjust the connections between the nodes. This is done every time there is an error. But when the utterance is correct no changes are made, or existing connections are reinforced. Thus, over time, with many stimulations and random adjustments where necessary, the network will settle into a state where it gets a high proportion of right answers. We will now have a system that has a cognitive capacity – that of recognizing the styles of various painters. Note that there is no reason to think that the system operates by anything like a rule. It may be true, for example, that the system has a special sensitivity to an image made up of small dots of paint, the characteristic pointillist style of Georges Seurat. But it would not be possible to identify some configuration of the network and say that it incorporates the rule, "If there are lots of dots, utter 'Seurat'", and the system may not be able to tell you that dots are characteristic of Seurat.

Connectionist models are now accepted as providing a far better understanding of brain function than the classical AI picture of rule-based expert systems. Connectionist systems, like brains, degrade gracefully. Nodes can be removed without total loss of function; loss of nodes leads to partial deterioration. Neuroscience shows us that the brain has a connectionist structure with neurons acting as nodes, electrically connected to other neurons, which is why connectionist systems are also called "neural nets". And human psychology shows the features we would expect of a connectionist rather than an expert system. Returning to the system for identifying painters, humans seem to be able to have this ability without following rules. A connoisseur might tell you with confidence and reliability that this painting is a genuine Pontormo and that the other is a fake without being able to put a verbal finger on what it is that makes a difference. This is not so readily explained if we assume that the connoisseur is operating with encoded rules for finding the difference.

Since connectionism was developed long after Kuhn wrote *The Structure of Scientific Revolutions* and the "Postscript 1969" and "Second Thoughts" paper of 1974, he was not to know of these developments. Nonetheless, it is just to credit Kuhn with the thought that not all human psychology can be explained in terms of rules and expert systems, and that in particular certain kinds of learning through concrete examples need not be seen in such terms. The most important claim in his book is that the activities of normal science are not best understood in terms of rules but rather in terms of learning through concrete exemplary puzzle-solutions. We thereby acquire what Kuhn calls "learned similarity relations". Apprehension of the existence of such a relation – directly seeing that two things are similar – is a non-intellectual exercise in that it does not require the exercise of reasoning, inference or the following of rules, either consciously or unconsciously. In connectionist terms, exemplars function as the prime source of input that trains the connectionist system. Exposing the system – the brain of the science student – to repeated exemplars, as well as seeking to adjust it when it goes wrong, via corrections to exercises, teaches it to recognize puzzle-solutions as being similar to the exemplars, the puzzle-solutions provided by and endorsed by the great scientists and the scientific community at large.

Because the process of recognizing similarity is non-intellectual, Kuhn has been accused of introducing "subjectivity and irrationality" into science. "Some readers," he wrote, "have felt that I was trying to make science rest on unanalyzable individual intuitions rather than on logic and law."[12] If I am right in taking Kuhn to be indicating features of human psychology that cognitive science has since been able to articulate and explain more fully, then such criticism is misplaced and he is justified in rejecting these charges. While it is true that use of "logic and law" is downplayed by Kuhn, it does not follow that the result is subjectivity or irrationality. It would be unfair to accuse connoisseurs of irrationality just because they cannot provide an explication of the discriminatory capacity they are able to exercise. Nor is the discrimination a merely subjective one. The connoisseur is sensitive to objective differences. There are at least two ways of verifying this. First, we may have an independent way of telling the real paintings from the fakes (e.g. chemical analysis of the paint). Secondly, if the connoisseur's attributions are consistent, and especially if other connoisseurs are able to make similar discriminations, it is a reasonable hypothesis that they are sensitive to some objective difference (that difference may not be the one we think it is, between fakes and genuine paintings but might be something different, e.g. between a late and an early style). It is something akin to this last point that Kuhn emphasizes. These learned discriminatory capacities, which we may reasonably call "intuitions", are not purely individual but are shared. They are learned by students in their training and are prerequisites for becoming a member of the professional group. Furthermore, our sensitivity to objective differences is explicable. Connectionism provides a well-confirmed model for such explanations, and even though Kuhn did not then know that, he himself was experimenting with computer simulations.[13] Ultimately, as Kuhn recognized, it is our neurophysiology that gives us our intuitions and gives them, in good cases, their reliability.[14]

Another way of understanding these shared intuitions is to regard them as embodying *tacit* knowledge. First, connoisseurs know when a painting is a fake and we would normally be happy in saying that they know the difference between the fakes and the genuine painting. But this difference that they know is not something they can usefully articulate. Kuhn contrasts tacit knowledge given by learning from exemplars with explicit knowledge of rules. But note that the following of rules need not be explicit. If by "explicit" rule-following one means conscious reference to the rule then much following of rules is not explicit. For the explicit following of a rule may become a matter of habit, at which point the rule is not being referred to in guiding behaviour. Driving on the legally required side of the road is like this. If there were rules in science we would expect that our following them would typically not be explicit. Kuhn's objection to rules would include rules that we are not explicitly following – rules are not to be extracted from the exemplars or from the behaviour of scientists trained by use of them.

Paradigms and disciplinary matrices

We shall now return to Kuhn's account of paradigms as exemplars and ask whether it was necessary to make the distinction between them and paradigms as disciplinary matrices. Certainly they are different and the exemplar is only one part of the matrix; other parts include shared symbolic generalizations, metaphysical commitments and values. The distinction suggests that the exemplar and the rest of the disciplinary matrix are independent – the same exemplar could be combined in a different matrix with different values, metaphysics and formal theories. If this is correct there will be an ambiguity in the phrase "paradigm-shift" and related concepts such as "revolution" and "normal science". For if there were a (revisionary) change to some non-exemplar part of a disciplinary matrix then we would have a shift of paradigm-as-matrix but no shift of paradigm-as-exemplar. It will be unclear what sorts of event are explanatory in Kuhn's theory. Are revolutionary scientific changes to be explained by changes in disciplinary matrices or by changes in exemplars?

If, on the other hand, exemplars play a privileged role within disciplinary matrices, rather as the chromosomes are a special part of a creature, we may be able to avoid such questions. The suggestion is that the exemplar might itself explain the other components of the matrix. If so, shifts in matrices would coincide with shifts in exemplars, thus avoiding problems of explanatory ambiguity. Furthermore, it would be clear that it is the exemplars that are explanatorily more fundamental. This would be a welcome conclusion since regarding matrices as explanatory of scientific change is dangerously close to tautology. For if the disciplinary matrix includes *all* shared commitments then a change, e.g. in a theory, would ipso facto just be a change in the disciplinary matrix and there would be no substantial sense in which the one explains the other.

How much reason is there to think that the exemplars determine the rest of the disciplinary matrix? Kuhn allows that not every aspect must be thus fixed. For metaphysical commitments may be varied or disagreed upon without change to, or dispute over, exemplars. Kuhn allows members of the same community to have different theories of matter, or, thinking of chemists in particular, to disagree about the existence of atoms, despite adherence to a common set of "tools" resulting from Dalton's atomic theory.[15]

It is of course open to us to regard such an admission as an acknowledgment that after all metaphysical beliefs do not, or do not always, form part of the disciplinary matrix. Yet Kuhn does want to regard metaphysical commitments as scientifically relevant; he explicitly mentions them as components of matrices in the 1969 Postscript. As Musgrave puts it, "Kuhn does not favour a return to the positivist view that such metaphysical disputes have no real influence on the practice of science proper".[16] An alternative understanding of what is going on is to see Kuhn as dropping the view that a community during normal science is characterized by *universal* consensus. Perhaps there is room for some level of disagreement that does not yet constitute a crisis. This fits with, and indeed exacerbates, the blurring between normal and revolutionary science discussed in the last chapter. In this view there may be more than one disciplinary matrix alive at a given time, whose differences correspond to differences, for example, in metaphysical commitments. For this to be normal science rather than pre-consensus or revolutionary science, these differences had better not be too great. We should expect there to be a high degree of similarity among the extant matrices in a given field (perhaps a family resemblance) or, more plausibly, there should be some common core (such as the same set of exemplars or symbolic generalizations). Either way it looks as if we must conclude that metaphysical beliefs are not always fixed by a commitment to exemplars. Is the same true of the other components of a disciplinary matrix?

It seems less easy to disconnect symbolic generalizations, heuristic models, and values from exemplars. In the case of symbolic generalizations the link looks to be particularly close, since exemplars are exemplary applications of such generalizations in puzzle-solving. If an exemplar consists of a solution to a puzzle concerning planetary motion and employs Newton's law of gravitation, a solution to another puzzle of planetary motion could not be regarded as similar to the first if it employed a different law of gravitation.

On the other hand, this last point should not be made too strongly. It may be that there is no consensus on the precise form of the equations or laws in a certain field. Perhaps there is room for some leeway with different detailed laws being used in puzzle-solving. This might have been the state of affairs in late-nineteenth-century radiation theory.[17] A consequence of this is that not all puzzle-solutions will be consistent with one another. Again there is a concern about the blurring of normal and revolutionary science. But this may not matter if we can see exemplars at work here. For example, if our exemplars consist of puzzle-solutions that employ different but formally quite similar equations (e.g. differing as regards values of constants) then we might regard as satisfactory a puzzle-solution that employs yet another variant as long as it does not differ too much. Of course, if all the exemplars given employ the same equation then it would be much more difficult to regard a proposed solution with a different equation as adequate but perhaps not impossible, if the solution is *de rigeur* in all other respects. This might allow us to consider Euler's and Clairaut's proposed alterations to Newton's laws as just about acceptable normal science.

Again we might want to regard this as an appropriate position for Kuhn's explanatory project. If sameness of exemplar *entailed* sameness of symbolic generalization, then again there is a danger of its not being possible to explain, in any interesting way, change in symbolic generalization by reference to change in exemplar. Put more positively, we do have an explanation of a commitment to symbolic generalizations in terms of a commitment to a shard set of exemplars, since similarity or identity of generalizations is a key factor in prompting us to see one puzzle-solution as being like another, and so like an exemplary puzzle-solution in particular.

Similar remarks may be made about the other components of the disciplinary matrix. How is it that a scientist acquires a set of values, and furthermore a sense of what counts as exemplifying some value (such as simplicity or fruitfulness) and a feeling for their relative importance? Here exemplars are a plausible answer, since it is difficult to say what simplicity consists in, even though it is easily spotted. As for face recognition a good explanation may be found in a neurophysiological habituation through exposure to exemplary cases.[18]

In summary, Kuhn's view is this. Scientists, typically as students, are exposed to exemplary problem solutions. Such exposure habituates scientists (in a way that may be investigated by psychologists and neurophysiologists) so that they directly perceive some new proposed solution as being similar to, or different from, the exemplars and thus accordingly a good or bad solution. Since the exemplars are common to members of the relevant community, members will have typically the same habituated dispositions – they will tend to agree or disagree on whether a problem solution is good or bad.

When I say that scientists directly perceive a problem as being similar to an exemplar I mean that they do not pick out certain features of each and then infer or calculate their degree of similarity according to some rule or algorithm. In Kuhn's view knowledge of similarity is unmediated. Nonetheless, certain features may be especially salient, features that when shared by two puzzle-solutions will dispose us more strongly to judge them to be similar. I have suggested that the best way to make sense of the relationship between exemplars and other aspects of community agreement (other components of the disciplinary matrix) is to think of those other aspects as being the salient features that propel us to make judgments of similarity. The case is similar to that of face recognition, where although there is no rule that allows us to recognize a face, there are nonetheless especially salient features of faces to which we are most sensitive when making judgments of recognition. Among problem solutions the most salient feature will be the symbolic generalizations employed. Also salient will be such features as simplicity, fruitfulness, and other "values". Still relevant but perhaps less salient will be metaphysical features of a solution.

Explaining scientific change

We are now in a position to reconstruct Kuhn's explanation of scientific change and so answer the questions posed at the outset of the chapter. There are two levels at which the explanation works: the social and the individual. We have just seen exemplars described at the individual level, providing scientists with a personal intuition for similarity. But they work at the social level too, for science is a collective activity, and as mentioned at the outset, *the* paradigm of a certain field is an object of consensus among its practitioners.

With our minds on the social level, let us recall the two questions posed for us by Hoyningen-Huene: (a) what is this consensus a consensus *about* and (b) what generates and maintains the consensus? We answered the first question straight away: the consensus is agreement on exemplary puzzle-solutions. Turning to (b), how does this agreement come about, and how does it persist? The question of persistence is easier to answer; the question of establishment I shall postpone for a few paragraphs. The key is the nature of scientific education and training. Learning to be a scientist requires training in the use of exemplars. Students spend much of their time engaged in various kinds of exercise. Some might be formal such as questions in textbooks; others may be practical such as laboratory exercises. In both cases there will be worked-out problems to follow, either fully worked problems in the textbook, or demonstrations in the laboratory carried out by an advanced graduate student or junior staff. Such exercises and practicals will typically not be common just to the students in that university but will be found throughout the world.

The uniformity in scientific education is remarkable, especially when compared with what is held in different places to be a satisfactory artistic, or even philosophical education. There are of course good professional reasons for this; failure to train students in standard techniques will prevent them from being accepted into the larger scientific community. Their puzzle-solutions will not be deemed adequate since they will not bear the appropriate resemblance to the exemplary solutions accepted by everyone else. Their papers will not be published – peer review in journals being a powerful force for uniformity. They will not land the jobs they desire and so will not be able to pass on their approach to others.

Let us return to the point of view of the individual, considered in more detail. In the training just described, young scientists are exposed to a variety of exemplary puzzle-solutions. These give them a sense, an intuition, of the similarity or dissimilarity of a proposed puzzle-solution to such exemplary solutions. This is what enables them to practise science. So exposure to solutions will give them a sense of what a good solution looks like. Note that the exposure is not passive but active, not just looking at exercises and demonstrations but practising them – which Kuhn likens to finger exercises for musicians. So the intuition gained will not merely be an ability to recognize a good solution when presented with one, but is also an ability to intuit, to think up, a good answer. Such an ability may be first exercised in "real" science when as graduate students they are put to work on some aspect of a larger research project. This is an apprenticeship, doing useful work, but work where the problems are set by someone else, the project leader, who will also be able to check the quality of the solutions presented. When the students have gained their PhDs they may be in a position to carry out independent work, which will often involve making applications to research councils, charitable trusts, and so forth. Now they need to be able to select good puzzles; their achievement and worth are measured not just by the quality of their puzzle-solutions but also by the quality of the questions they have sought to answer. Again the exemplars do the work here. For an exemplar must consist of both a puzzle and its solution. And so exposure to exemplars will give students not only a sense of what good solutions are but also an ability to sense the similarity of a possible research puzzle to an exemplary puzzle. This may be more difficult for young scientists, for as students they not only saw puzzle-solutions but practised generating them. But students are only rarely asked to think up new puzzles themselves, and so their ability to recognize worthy puzzles may be better trained than their ability to actually conceive them.

Thus we can see conservative forces operating in two ways: one at the individual level and another at the social level. At the individual level the nature of the exemplar in generating a sense of similarity – along with a sense of the positive value of that similarity, and in forging an ability to generate what is similar to the exemplar – will result in the individual being disposed to produce conservative, non-revisionary solutions that look like the exemplar, discarding attempted puzzle-solutions that are unlike it. At the social level such exemplars are shared by the whole of the relevant scientific community; indeed the most successful scientists – those who have most influence and power – will be the ones who have created the exemplars in the first place. A failure to acquire the appropriate intuitions of similarity and corresponding dispositions will lead to an individual being weeded out of the scientific community at some stage, whether as an undergraduate, at the PhD level, when seeking funding for a project, or when trying to publish the results.

The exemplar fulfils, therefore, both for the individual and for the scientific community, three of the four functions listed on pp. 68–9: puzzle identification, solution identification and research assessment. (The fourth, semantic, function is rather different – I shall discuss it in Chapter 5.) As long as the consensus holds, science governed by an agreed set of exemplars is inherently conservative, avoiding novelty, outlawing deviation from existing puzzle-solutions, discouraging unusual research projects. Since non-exemplar features of the disciplinary matrix – symbolic generalizations, metaphysical beliefs, values, heuristic models – are the results of agreement on the exemplars, and because a satisfactory solution is one that resembles exemplary solutions, it is these features of solutions, being especially salient in resemblance, that are the most rigidly conserved.

Consensus and conformity cannot be maintained under just any circumstance, for the aim of each scientist is to provide puzzle-solutions. In normal science it is adherence to the pattern set by our exemplars that plays the major role in generating puzzle-solutions, in particular solutions acceptable to our colleagues. Indeed, exemplars not only give us a sense of similarity but provide us with expectations – expectations that puzzles similar to those solved by the exemplars will be solved by solutions themselves similar to the exemplar solutions. If we repeatedly fail to find solutions and instead generate anomalies, then something has gone wrong. What has gone wrong may be a fault in the individuals – they are just not good scientists, training with the exemplars hasn't given them the appropriate dispositions required to spot puzzle-solutions. But if sufficiently many colleagues are in the same boat a common feeling might emerge that it is the fault of current science, the theories, techniques, and so on in which they were trained. Just as the sense of similarity and the expectation of future success are reinforced by repeated past successes that resemble the exemplar, repeated failure to find adequate solutions will weaken the strength of expectation. And when the strength of expectation is weakened to such an extent that a good proportion of the community has begun to doubt whether solutions to anomalous puzzles will be forthcoming that resemble the exemplary solutions, then consensus breaks down – the community is in crisis.

We need now to see how the concept of paradigm-as-exemplar may help us explain not only how normal science may persist – and break down – but also how it may be re-established. Some of Kuhn's critics have seen this question as reinforcing their view that theory-change for Kuhn is irrational, a matter of mob rule. In this interpretation the rejection and replacement of a paradigm is wholesale. And once a paradigm is rejected there is then no basis for choosing a new one. If there is no accepted paradigm how is the new signal achievement around which a new consensus is to crystallize recognized as such? It is tempting to elevate Kuhn's few isolated hints about the possible significance of external social and political conditions into the claim that it is these factors and these alone that determine the choice of new paradigm.

Kuhn's view is more sophisticated than this overly simple interpretation suggests. Kuhn does in fact allow the old paradigm to play a role is determining its successor, the latter must solve the problems of the former and also maintain many of its successes; it must bear *some* resemblance to it. In a crisis members of the community may be willing to contemplate puzzle-solutions that bear far *less* resemblance to the exemplars than would be expected of a normal puzzle-solution. There is a trade-off here – either accept this unusual solution, despite its divergence from the exemplars, or live with the unsolved anomalies. There is no compulsion to accept the trade; the solution is neither demanded by logic, nor is it compellingly similar to the exemplars (rather the opposite). But if the trade is accepted, then implicitly one can no longer regard similarity to the old exemplars as definitive of a good puzzle-solution. Instead then, the new "unusual" solution becomes a standard by which, for those that accept it, yet later puzzle-solutions (and puzzles themselves) are judged.[19] It is thus that we have a revolution, the acceptance of a non-standard puzzle-solution, and the commencement of a new period of normal science employing that solution as a new exemplar.

Problems with Kuhn's explanation

With this sketch of Kuhn's explanation of the cyclical pattern of scientific development in place, we may now ask how good an explanation is it? While it does make an important and innovative contribution towards a full explanation, there are a number of reasons for doubting it to be completely adequate as it stands.

Plurality of functions
Some of the problems with the exemplar explanation focus on the exemplar concept itself. We examined four functions of exemplars. Three of these belong together – puzzle identification, solution identification and research assessment which I will call the "puzzle-solving" functions. The fourth function – the semantic function of exemplars – seems rather distinct, having little obviously to do with the puzzle-solving practice of normal science. While we might expect an exemplar to fulfil all the puzzle-solving functions simultaneously, it is not at all clear that an exemplar might also fulfil the function of giving meaning to empirical concepts. Concepts may well indeed be learned through use of exemplars; the question is whether the same exemplars may be used in providing standards for puzzle-solving and in empirical concept formation. Empirical concepts are those that may be employed non-inferentially through perceptual contact with the situation to which the concept is to be applied; empirical concepts are perceptually triggered. Classic cases are colour concepts, although Kuhn is keen to extend the range of concepts considered empirical well beyond the like of these. Whatever that range may be, all we need to notice is that in order to be used in empirical concept formation an exemplar must be a perceptual entity (an object, an experimental situation etc.). Only things that are perceptual can be used as exemplars in training someone to use a perceptually trig-gered concept.

Puzzle-solutions typically are simply not perceptible in this way; a puzzle-solution will be a chapter of a book or a scientific paper. It may involve perceptible things such as diagrams, or describe perceptible things such as experimental arrangements. But the perceptual things are not themselves enough to constitute the puzzle-solution. Hence not everything that is an exemplar-as-empirical-concept-former is an exemplar-as-puzzle-solution. And since quite a large range of puzzle-solutions do not involve diagrams, experiments or anything else perceptible beyond the printed work, not all exemplars-as-puzzle-solutions are exemplars-as-concept-formers.

Consequently it is misleading to think that exemplars have both the function of evaluation of puzzle questions and puzzle-solutions *and* the function of concept formation. Rather, there are two different sorts of exemplar, one sort for empirical concept formation and another for puzzle evaluation. This does not deny that one thing might belong to both sorts – a puzzle-solution may have a perceptible element although even this might be better described as an exemplar of one sort that has as a component an exemplar of the other sort.

This criticism is not deeply damaging to Kuhn's explanatory project since we can ascribe the power to explain scientific development just to the exemplars with puzzle evaluating functions, and leave aside the concept-forming exemplars as a separate matter. From now on we can regard exemplars, in the context of the explanatory project, as having (at most) the three puzzle-solving functions.

There are nonetheless important consequences for Kuhn's overall view that deserve our notice. Divorcing the concept formation function and ascribing it to another kind of exemplar means that we have to give up an attractive but ultimately spurious unity in the Kuhnian picture. The seeing of a gestalt is the perception of a significant form, and so, if a revolutionary change in exemplar is a change in a perceptible item, then a revolution may be described as literally involving a shift in gestalt. But if we remove the perceptual, i.e. empirical function from revolution-explaining exemplars, there is no reason to think that a change in the latter involves any actual change in gestalt as opposed to simply an analogy with such a change.

Furthermore, if all the functions were combined in the same exemplars a change in exemplar would immediately suggest a change in concepts. Then arguing that revolutions bring about conceptual incommensurability would be fairly straightforward. Separating the functions makes the case rather less simple. These comments reflect only on the supposed function of exemplars in fixing empirical concepts – for all that has been said they might have a role in fixing theoretical concepts. This we shall examine in Chapter 5.

Explaining puzzle selection
In the foregoing I have accepted that we might well expect an exemplar to possess all the puzzle-solving functions together. In this section I want to question whether an exemplary puzzle-solution is a good explainer of puzzle selection. In the descriptive project Kuhn talks of normal science as being a "mopping-up operation" – filling in gaps left by earlier scientists and, in particular, by the revolutionary pioneer. Some such mopping-up activities can be seen as resembling earlier ones; scientists who sought to get an accurate fit between Newton's laws and observed planetary motion were doing just as Newton did, their projects and aims being very similar to his.

But not all normal scientific projects have exemplary predecessors. Cavendish's measurement of the gravitational constant G had no predecessors. Yet Kuhn rightly regards measurement of key constants as normal science – Cavendish's was certainly not a revisionary project and was very much a matter of filling a gap in the Newtonian picture. Yet his measurement could not have been legitimated by a perceived similarity with some earlier activity. What made it worthwhile was that Newton's theory tells us that there is a constant without telling us what its value is.

Of course, insofar as the exemplar explains a commitment to a certain symbolic generalization it may also explain an interest in the values of the constant term contained within it. This may well be sufficient to allow a single exemplar both of the explanatory functions being discussed. But note that it will not (always) be the case that in explaining the selection of a research problem the exemplar will do that by being *exemplary* of that kind of research problem. Selecting this research problem will be a more intellectual activity than can be explained in terms of apprehending a similarity to an exemplar.

Research assessment
A corresponding doubt is raised over the functions of finding and evaluating puzzle-solutions. If exemplars cannot exemplify all possible choices of puzzle, it follows that they cannot always explain how we find puzzle-solutions, nor how we assess them. Because Cavendish's puzzle was like no previous puzzle, his solution was not like any existing exemplary solution.

Symbolic generalizations and theory
The foregoing concerns with explanation by exemplars highlight the fact that there is a downplaying of *theory* in the account of exemplars

that is in tension with the emphasis on *theory* change in the descriptive project. Revolutions are primarily changes in theoretical beliefs. Yet Kuhn gives no clear account of why theories should change when one exemplar replaces another. The nearest we get to theory is the mention of symbolic generalizations as one component of the disciplinary matrix.[20] I explained that the various components of the disciplinary matrix, other than the exemplar itself, are dimensions of salience relevant to judging the similarity of proposed puzzle-solutions to the exemplary solutions. Since exemplars may differ along other dimensions (e.g. in terms of values or metaphysical commitments etc.) we might expect to be able to replace one exemplar with a contrasting one that nonetheless had the *same* symbolic generalizations and theoretical commitments. It ought in principle be possible to replace Newton's theory of gravitation-plus-action-at-a-distance by the same theory-plus-no-action-at-a-distance and regard these as different exemplars. The replacement of one by the other would be a revolution, just as the replacement of Newton's equations by Einstein's was a revolution. Similarly we ought to be able to conceive of revolutions that consist just in the decision to regard success in explaining observations as less important and simplicity or elegance as more important, while keeping the same symbolic generalizations. But exemplar shifts that are shifts just of metaphysics or of values receive little discussion and the consequences for the history of science are not spelled out. The reason for this may well be that shifts of this sort rarely take place. Changes in metaphysical and value preferences do occur but this is typically because they accompany the theoretical changes that are the focus of attention.

It seems therefore that the theoretical dimension to exemplars must be especially significant. As already suggested, similarity with respect to symbolic generalizations must be regarded as an especially salient feature when assessing resemblance to exemplars. Correspondingly, replacement of an exemplar by one that differs theoretically will be a highly significant shift with more substantial historical consequences. This, it must be said, is an elaboration of Kuhn's explanation, and not something he addresses. It is a plausible elaboration, for it is clear that in solving puzzles it is the symbolic generalizations and other theoretical claims that do the lion's share of the work.

It is perhaps more plausible that our sense of simplicity and our senses for other values are given to us by habituation with exemplars. It is less plausible that our sense of similarity with respect to theory and symbolic generalization is acquired in the same way. We have already discussed whether similarity is the appropriate concept here, whether puzzle-solutions with slightly different symbolic generalizations could be acceptable. If different generalizations are not acceptable, then the idea of a "learned similarity relation" seems out of place here – straightforward identity is enough. If differences are permitted, it is unclear how the exemplars give us a sense of what is similar and what is different since all the exemplars (such as Newton's problem solutions in *Principia Mathematica*) use the same symbolic generalizations. The problem is analogous to the difficulty created by trying to teach someone the word "red" by repeated use of one and the same shade of red.

A deeper objection concerns Kuhn's downplaying of the use of rational inference in assessing the goodness of a puzzle-solution and its similarity to an exemplar. This seems parallel to the downplaying of the theoretical/symbolic dimension of exemplars. Kuhn points out that a law may be expressed in a variety of different forms. We do not always get to use Newton's third law in the form $F = ma$; sometimes, for free fall, we have $mg = d^2s/dt^2$; or for a simple pendulum $mg \sin\theta = -ml(d^2\theta/dt^2)$; or, for a pair of interacting harmonic oscillators, $m_1(d^2s_1/dt^2)+k_1s_1 = k_2(s_2 - s_1 + d)$, and so on.[21] How does the student learn, asks Kuhn, to find the right version? Kuhn's answer is the by now familiar one of learning with exemplars: "The law-sketch, say $F = ma$, has functioned as a tool, informing the student what similarities to look for, signaling the gestalt in which the situation is to be seen. The resultant ability to see a variety of situations as like each other, as subjects for $F = ma$ or some other symbolic generalization, is, I think, the main thing a student acquires by doing exemplary problems, whether with a pencil and paper or in a well-designed laboratory".[22] Kuhn may well be right that a feel for what might be a good way to approach a problem and which version of the law will be most helpful is acquired in just this way. But that does not mean that the adequacy of student's choice or his/her problem answer is to be judged solely by using the same sense. We can prove mathematically that the variants Kuhn lists all follow from the basic formulation $F = ma$ – we do not have merely to intuit their similarity. Indeed students may well be expected to show such a derivation in their problem solution, and the quality of their answers will be assessed using standard tools of logic and mathematics, whose application may be mechanical. Seeing right variant is often a matter of knowing what assumptions are needed in order to derive that variant from $F = ma$. This is analogous to the case of a chess grandmaster whose powers of pattern recognition allow him to see the current state of the board as ripe for a forced mate, but if a certain move will lead to a forced mate in four moves then that can be proven without recourse to that pattern recognition power.

As already explained, my own view is that making judgments on the basis of learned similarity relations is perfectly compatible with such judgments being classified as rational, or at least with their not being classified as irrational. Kuhn has done us a service in drawing our attention to the existence of such relations in scientific judgment. It is however an exaggeration to claim that such relations are *sufficient* to account for all judgment in science. While Kuhn does not explicitly make such a claim the failure to discuss any other source of judgment, except rules, which are dismissed, suggests that this is what he thinks. A far more plausible picture of scientific judgment would see it as involving both intuition (including judgments of learned similarity) and reflective reason, which presumably is the natural way to think of other kinds of puzzle-solving (chess problems, crossword puzzles, poker playing etc.). In particular an account of the sensitivity of judgment to theoretical or symbolic components of exemplars and puzzles will have to take due account of our reflective capacities as well as the intuitive ones. How this might be done in a broadly Kuhnian framework I shall discuss later.

Quantity of exemplars

A chess problem may be such that only an experienced player can immediately see the solution, while a less advanced player may be able to work out the answer. Acquisition of a similarity relation typically requires repeated exposure to exemplars. In simple cases, such as the learning of colour names, perhaps only a few samples are required. But seeing more complicated patterns and resemblances will typically require repeated exposure to a variety of different instances. Given the complexity of puzzle-solutions one might think that to acquire the appropriate sense of similarity a student would have to be exposed to a very large number of exemplary solutions. Then we may ask, do we think that this is in fact the case? Perhaps one may argue that the number of exercises and laboratory demonstrations that a student experiences is sufficient. But note that most of these are not taken from the work of the revolutionary pioneer but were devised in the process of normal science. So we may ask, where did the authors of these puzzles and solutions get their learned similarity relations from? And so, more pressingly, we may ask of scientists working in the early phase of a new period of normal science, who will have at hand far fewer exemplars than a student later in that science's history, is it psychologically plausible that they have had enough exposure to exemplars to acquire the appropriate similarity relations? To answer this question would require collaboration between psychologists and historians of science, so I shall not attempt an answer. But two comments are pertinent. First, as long as the question is open, Kuhn's account must be in some doubt. Secondly, the problem of psychological plausibility would be far less pressing if we allowed a role for reflective capacities as well.

Gradual or radical shifts?

Talk of similarity relations suggests a continuum of greater or lesser degrees of similarity, which is why the question of small shifts in symbolic generalizations could arise. If so there ought to be potential puzzle-solutions that are on the borderline between similarity and dissimilarity to the exemplar. Imagine that there is a puzzle-solution just beyond the borderline which therefore is not generally acceptable, and then imagine that the puzzle to which it is proposed as a solution has not been solved by any solution more similar to the exemplar. Then this borderline solution might well, in due course, be accepted on the grounds that this requires no very great liberalization of the standards of similarity. Such a move would not involve changing the exemplar although the newly accepted solution would have changed the *use* of the exemplar. Presumably this should reveal itself as a certain sort of episode in the history of science – something short of a full revolution, but mildly radical nonetheless. While we saw that Kuhn says things that are evidence that he partially recognizes a continuum between normal science and revolution, he gives little systematic treatment of intermediate cases or their causes.

Acceptance of new exemplars

I have just suggested that, if Kuhn is right about the importance and role of exemplars, gradual shifts in the use of exemplars might be possible. Here I wish to ask how replacement of exemplars could be rationally possible. An analogy to the gradual adjustment case might be this. We are given the task of decorating a room with some colour that is similar to certain exemplars of colour, all of which are shades of red.

After a failure to find any red paint we might be willing to settle on orange paint as a compromise. Our existing exemplars and the use previously made of them suggest that this is not quite the solution we wanted, but the same exemplars suggest that it is only a small liberalization to allow the orange item. The case of the revolutionary replacement of exemplars would be like a decision to accept green paint. Such a decision is in no way legitimated by existing usage – it is clearly not even an approximation to an adequate answer. If this were a fair analogy, then Kuhn's critics would be justified in the claims that his account makes revolutionary change irrational. Of course, as we have seen, Kuhn denies such charges, replying that revolutionary change can be rationally supported and argued for; it just is not rationally *compelled*. But if there is room for rational argument then Kuhn needs to provide some mechanism for this in his explanatory apparatus. And that is just what non-reflective judgments of similarity to exemplars seem to lack. In particular Kuhn holds that a new paradigm must solve some outstanding problems and preserve some of the problem solving ability of its predecessors.[23] This suggests an element of problem solving that is independent of similarity to established exemplars, in which case exemplars cannot be the whole story about what makes a supposed problem solution acceptable. Conversely, if what counts as problem solving *is* defined by exemplars alone, it is difficult to see how revolutions as Kuhn described them could take place. For if the new candidates are genuinely different then they cannot be seen as solving any existing problems, precisely because solving problems requires similarity to existing exemplars. Trivially the new candidate will be a good problem-solver by the standard it sets, but this provides no explanation of why anyone should want to adopt it while making no room for the partial continuation of problem solving from the previous paradigm.

Perhaps the colour analogy overstated the difference between the new and the old; perhaps paradigm shifts *are* like moving from red to orange or yellow, where some sense of continuity is available. Even so, the complaint remains that Kuhn's notion of exemplar is simply too thin to do justice to the complexities uncovered in the detailed description of scientific change.

Reasoning and intuition within paradigms

Logical empiricism embodies the thought that acceptable scientific development could be fully explained solely using rules of inference, and *a priori* justifiable rules at that. The logic of deductive inference seemed to supply the rules for the non-empirical sciences, and so Carnap, for one, sought to extend deductive logic with an inductive logic so as to provide the rules for the natural sciences. It is because he believed that no such rules could be found that Popper denied that that we have any reason to believe a theoretical claim in science. Kuhn's radical break with logical empiricism was to deny this role to rules, replacing them with similarity relations learned from exemplars. The model for the learning of similarity relations is the acquisition of perceptual powers of pattern recognition. Exercising such a power is something that typically occurs in a single psychological moment of recognizing, as when one sees the duck-rabbit as a rabbit, or perceives one rabbit as being like another. Such acts do not employ reasoning or reflection; indeed they seem to leave little room for it. And so Kuhn's preferred model for explaining our puzzle solving and other normal science practices is one that seems to be in tension with the undeniable empirical fact that scientific thought is deeply reflective. Appreciating that one puzzle-solution is like another is not like suddenly seeing a face or figure in a gestalt picture. Instead it involves a sequence of non-perceptual thoughts that relate to one another in a structured fashion.

Kuhn may be forgiven for wanting to draw his contrast with the logical empiricist emphasis on rules as strongly as possible. While I would not accuse Kuhn of being ignorant of platitudinous facts about scientific thinking, I do suggest that that in appealing to a model that has no need of rules, he makes it unclear how reflective judgment can play a role. This lack of discussion of how reflection fits in means that Kuhn's story is sorely incomplete. So our next question must be whether there is room for additional claims about the role of reflection. If not, Kuhn's account if implausible; if there is, it stands in need of supplementation.

There are two ways in which one might fuse reflection and intuition in a broadly Kuhnian way. The first suggestion is that the intuition of similarity is preceded by reflection. The rough idea would be this: science is full of reflection and reasoning, but these are not sufficient to fix an answer to the question: how good is this hypothesis? Reflection can do things such as determining whether the evidence might be explained by the hypothesis, whether a certain claim is statistically likely, whether the calculations of the projected path of a comet have been carried out correctly. These things are reflected upon and determined before we can answer the question of the quality of the hypothesis. But there is no rule that takes us from

91

our answers to the reflective questions to an answer to the quality question. There comes a point at which we must say something like "All things considered, I judge this to be a plausible but not yet fully convincing hypothesis", which contrasts with, for example, a mathematical case where one may say something like, "Having seen the proof, I see that it follows logically that the theorem must be true".

It is certainly plausible that our judgments of the quality of a hypothesis are the products of intuition fed by reflection. My suspicion is that this is how Kuhn saw the relationship. Reflection on both an exemplar and a putative puzzle-solution is what allows us to intuit their degree of similarity. In this view reflection in the scientific case is analogous to the careful visual inspection in the perceptual case. No one should think that, in difficult cases, a connoisseur's intuition can be elicited by a mere glance.

Nonetheless, even though it accounts for the existence and importance of reflective judgment in science, this view still has problems. Even if it is plausible that judgments of quality rest upon a final act of intuition, it is not at all clear that this power of intuition is best explained by our having learned a similarity relation from exemplary problem solutions. Many of the problems raised above still persist. For Kuhn exemplars are the products of the revolutionary pioneer; are there really enough exemplars to forge an intuitive sense of similarity? And how do we explain our ability to appreciate original puzzles and their solutions, such as Cavendish's, by reference to any learned similarity relation?

There is another way in which intuition and reflection might work together. It is not obvious that reflection requires rule-following; this fact and the possibility just canvassed that intuition may be fed by reflection together may suggest that there is not so great an opposition between intuition and reflection as at first appeared. A scientific theory is, as Darwin described his *Origin of Species*, "one long argument" and a puzzle-solution in normal science is merely a shorter argument. But rarely are such arguments anything like proofs found in mathematics or logic, even if all the scientists' stated premises are granted. The question then is, what, in addition to the mathematics and logic, makes us accept some arguments, or steps in arguments, and not others? It is plausible to suppose that as a matter of fact we do accept arguments or argument steps that have features similar to those found in theories we already accept as successful or were taught as exemplars of good scientific reasoning. This may be part of the significance of Kuhn's notion of paradigm-as-exemplary-

puzzle-solution. The deductive logic employed in scientific arguments has little changed over the centuries (even if we have learned
how to formalize it); the mathematics has developed, but largely in a
cumulative fashion. But the interesting feature of exemplars, according to Kuhn, is that changes in our stock of exemplars are not typically cumulative but are revisionary. Hence our assessment of the
quality of a proposed puzzle-solution may undergo a shift. In extreme
cases this shift may dispose us to reject arguments we previously
accepted and vice versa.

One reason that this might be possible is that because in addition
to the overtly stated premises and principles employed there are
unstated assumptions and implicit premises. One role of exemplars
may be to supply and inculcate these. Furthermore, if we think of
tacit assumptions as those that could be stated explicitly if only
someone, perhaps some later historian, were to try hard enough,
then it may be that even these – along with logic, mathematics and
the overtly stated premises – are insufficient to explain why someone
employs the argument they do and why someone else accepts it. Let
us think about logic for a moment. There are certain recurring simple
patterns in our arguments, patterns that are pretty well universal.
Even so, it took some hard work by Aristotle and his medieval successors to identify an important set of these and to show that they bear a
systematic relationship to one another. Frege extended the range of
patterns covered by "formal logic" while metalogical studies showed
why, given the concepts of truth and falsity, these patterns should
arise, if people argue with the aim of avoiding falsehood. It is apparent then that people have an ability to learn to use and recognize
patterns in reasoning independently of being able to articulate those
patterns.[24] There may be other patterns of reasoning that on the one
hand are not as ubiquitous as logic but are not to be thought of as
assumptions.[25] So exemplars may inculcate not only implicit
assumptions and beliefs but also patterns and habits of thought. And
hence a puzzle-solution from another paradigm may seem unfamiliar, not just because it makes different assumptions but also because
it incorporates different habits of inference.

Thinking of things in this way may help resolve a puzzle in Kuhn's
characterization of the intellectual basis of revolutionary change in
paradigms. Kuhn repeatedly insists that he does not regard such shifts
as irrational, let alone a matter of "mob-rule". Nor is his account
a manifestation of relativism. Rather, what is characteristic of
revolutionary change is that "the force of logic and observation cannot

in principle be compelling";[26] revolutionary theories are not susceptible to proof. This suggests that Kuhn thinks that logic and observation
can be compelling in normal science; that proposed puzzle-solutions
can be proved to be genuine solutions. Yet in the discussion of normal
science, notions of logic and proof do not get mentioned. The determi-
nant of puzzle choice is similarity to an exemplar. Furthermore, the
sort of learned similarity relation that puzzle choice is supposed to
exemplify seems to be explicitly contrasted with logical empiricist
notions of proof, or deducing theoretical conclusions (or their probabili-
ties) from observations, and so on. If normal scientific advances do
not employ proofs, then what is the intellectual difference between
accepting a puzzle-solution and a revolutionary theory?

If we were to concentrate on the word "compelling" in what Kuhn
says, it might be sufficient simply to say that because in normal
science good solutions are far more similar to existing exemplars
than are revolutionary proposals, the former can be psychologically
compelling in a way that the latter are not, just as one mallard is
compellingly similar to another in a way that a goose is not. However,
we may do more justice to Kuhn's mention of logic and proof if we
accept the proposals made above. Grasping a likeness between exemplar and puzzle-solution is not a matter of a one-step sensing of
similarity. Rather it is a multi-step process, involving a number of
manifestations of learned similarity, which altogether may count as
a proof, in many cases if not always in all. When it comes to a revolutionary theory it may be that not all the steps in the argument
exemplify such learned similarities, or that the overall argument
does not engender a sense of familiar reliability because existing
exemplars do not readily cause us to identify the new theory as a good
explanation of the evidence.

If we think of intuition as the power to make judgments without
following rules, then the core of Kuhn's contribution to the philosophical question of theory choice is to point out, in opposition to the
logical empiricists, the significance of intuition. However, it is doubtful whether Kuhn was right about the origin and functioning of this
intuition. He took its source to be an exemplary achievement by a
revolutionary pioneer. Furthermore, he was inclined to think of the
operation of intuition as something unreflective, first, because his
understanding of intuition is influenced by rather simple perceptual
cases where reflection is largely absent, and secondly, because reflection seems to suggest the operation of rules and so would seem to
contrast with intuition.

A more plausible account of the nature and function of intuition in science ought to have a more sophisticated view of its relation to, and role in, our reflective capacities. That account ought then also provide a more detailed picture of how scientists' experiences might mould their powers of intuition. The existence of pioneering exemplars that generate an intuition of overall similarity is too simplistic an explanation. For one thing it ignores innate intuitive powers. But even if we restrict our attention to learned capacities, we need more for the explanation to be satisfactory. We need to see that parts of exemplars might give us intuitions about parts of puzzles and their solutions. One argument is not intuited directly as being like another, but indirectly, in virtue of intuited similarity of steps in the reasoning. This not only allows for the structured, often sequential nature of reasoning, but also permits a worthy puzzle and its solution not to resemble an exemplary puzzle and solution overall. For if the parts of a piece of reasoning are familiar we may judge the whole to be acceptable even if quite novel, just as we can understand a new sentence on the basis of understanding the constituent words and recognizing simple structural combinations.

Conclusion

I mentioned that Kuhn believed paradigms-as-exemplars to be the "most novel and least understood" aspect of the thesis of *The Structure of Scientific Revolutions*, and having looked at Kuhn's work with the benefit of hindsight I am sure that he was right. His own imprecision in the use of the paradigm concept was partly responsible for that lack of comprehension – and indeed for subsequent misuse of the concept. Kuhn spent much of the rest of his career trying to emphasize the idea of an exemplar that forges a learned sense of similarity. But there are two other reasons why this aspect of his thinking was not sufficiently appreciated.

First, Kuhn failed to work the idea out in sufficient detail. It is one thing to provide a new model of cognition – intuitive pattern recognition – in opposition to another model – conscious following of methodological and other rules. It is another to show that this adequately answers to the various detailed facts. In particular there is a failure to relate the model to the structural nature of almost any piece of scientific thinking. Part of the reason for failing to develop the idea is that he tried to use it for disparate purposes. Initially Kuhn employed the exemplar idea primarily in two roles, in explaining certain features of perception and in explaining scientific change. Later he used the exemplars more in the explanation of learning meanings and in relation to the thesis of incommensurability. Kuhn himself is unsure about the relationship between the various roles: is it that they are aspects of one and the same phenomenon (one and the same because characteristic of the mind); or are the roles just analogous to one another?[27] Had Kuhn sought to be clear about this relationship, he may have been forced to become aware of the detailed problems that face his somewhat sketchy account of the role of exemplars in explaining the history of science.

Secondly, Kuhn's new picture lacked, at that time, sufficient theoretical support. Kuhn said that it is the subject's neurophysiology that is ultimately responsible for giving him/her learned similarity relations. But when he was writing there was no research to show how the brain might do this. Now we can see that connectionism provides exactly the mechanism Kuhn was looking for. In partial mitigation of the criticism that I have levelled at Kuhn in the last paragraph and in much of later part of this chapter, it should be noted that contemporary connectionists face exactly the same difficulties. Some critics of the connectionist model of the mind, such as Jerry Fodor, argue that connectionism will never account for the structured features of thought. Even those researchers who reject Fodor's arguments concede that sequential reasoning represents a challenge to connectionism that is only beginning to be met. It is reasonable to expect that when connectionists are able to provide a fuller and more detailed account of thinking, Kuhn's view will be vindicated, and appropriate answers to the questions I have put to him will become apparent. Unfortunately Kuhn himself abandoned interest in the support that empirical studies could lend to his explanatory theory. As his career progressed Kuhn's interests and approach became, in general, more philosophical and sceptical and less historical or scientific.

Chapter 4

Perception and world change

Empiricism, perception and observation

In the last chapter we looked at Kuhn's notion of exemplar, which he employed in the explanation both of normal science and of perception. A scientific revolution requires the replacement of one exemplar by another. This is liable to involve, according to Kuhn, not only change in theoretical belief but also in perceptual experience. In this chapter I shall examine Kuhn's views on perception, most especially concerning the relationship between perception and theory. In the light of this I shall examine the well-known thesis that paradigm changes bring with them changes in the world

Kuhn's views on perception must be understood against the background of standard logical empiricist assumptions about perception, observation and their significance for science:[1]

  (i) *the observational basis* Judgment in science is founded on the observations we make (e.g. hypotheses are tested against observational evidence);
 (ii) *the experiential basis* Observations are reports of perceptual experiences;
(iii) *internalism* The content of a perceptual experience is immediately knowable to its possessor – if one has an experience, one can know straightforwardly that one has it and what it is;
(iv) *independence* Perceptual experiences are raw data – they do not imply or depend upon any judgment or related mental state.[2]

The empiricist account of observation seeks to ensure that observation is an epistemologically secure foundation for scientific inference, and in particular that observation provides a shared base for theory choice. Assumption (i) says that observation is the basic level of judgment in science. Scientific judgment is fallible – it might be true or false – so in order to provide a secure foundation for science we need to find some basis for observational judgment that does not itself involve any judgment. Assumption (ii) says that perceptual experience plays this role. Experiences are not judgments. Our basic judgments, observations, are judgments about the content of perceptual experience. The internalist assumption (iii) ensures that at least observational judgments are not fallible, and so provide a secure basis for scientific inference. The independence assumption (iv) ensures that the foundational role of perceptual experience is not compromised. Not only are perceptual experience not judgments, they are not dependent on judgments either. Furthermore, since perceptual experience and hence observational judgment are independent of any other judgment, observation is an independent court that may decide between conflicting scientific judgments of a non-basic kind.

Kuhn's attack on empiricist perceptual doctrines is directed at the assumption of independence (iv). While Kuhn rejects the idea that observation provides an infallibly secure foundation for scientific inference, he is more interested in undermining the idea that observation is the independent arbiter between conflicting scientific beliefs and theories. He does not deny that observation can play this role in normal science, but he does want to deny that it can have this status when the disagreement is in choice of paradigm theory. For what we may observe, he claims, is itself affected by one's paradigm – independence does not hold. Consequently, proponents of competing paradigms, such as Lavoisier and Priestley, cannot resolve their differences simply by, for example, carrying out a crucial experiment whose outcome is an observation that both are agreed will decide between them. We have seen already that Kuhn rejects the idea that there are any rules of theory choice that compel the adoption of a particular paradigm. Rejecting the independence assumption bolsters this claim. The empiricist view is that observation provides the raw material for such rules to process. Kuhn is saying, in effect, that even if there were such rules, there are no paradigm-independent observations for such rules to get to work on. Overall, the view can be summarized as: since observation is not paradigm-independent, observation is not, *pace* the empiricists, a common measure of the quality of theories. The view is thus an aspect of Kuhn's more general claim that paradigm theories are incommensurable.

One of the most important sources for Kuhn's assault on the independence assumption about perception was Norwood Russell Hanson's *Patterns of Discovery*.[3] Since Kuhn follows Hanson very closely, and because Hanson's ideas on perception are expressed more systematically than Kuhn's, it will help to start by looking in some detail at Hanson's account of perception and observation.

Hanson on perception and observation

In the first chapter of *Patterns of Discovery* Hanson argues that the nature of observation is far more complicated than is credited by the logical empiricists. In particular, there is an important connection between what we observe and our background beliefs and past experience. While Hanson does succeed in undermining the logical empiricist position, it is also true that he imports a number of confusions that affect not only his positive account of observation, but those also of subsequent commentators, including Kuhn.

Empiricist assumptions (i), (ii) and (iv) together argue that since observation is the foundation of scientific judgment, observation itself must not presuppose any prior judgment, knowledge or belief. We should expect therefore that observation should be the same for all observers, as long as they are similarly placed and have the same senses, functioning equally well. Hanson attacks this feature of the empiricist view by arguing that similarly situated observers do *not* make identical observations; instead what they observe depends on what they know or believe. In particular observation depends on theoretical belief, i.e. observation is *theory-dependent*.

Hanson's argument proceeds primarily by offering examples of cases where we would (he claims) say that one observer observes something that the other does not (and sometimes vice versa). The cases are varied and require, as we shall see, separate discussion. A relatively simple case involves an experienced scientist and school-boy looking at the same piece of laboratory equipment. According to Hanson, the former sees an X-ray tube while the latter does not. In a more sophisticated case, two microbiologists provide different descriptions when looking at the same slide. One sees a certain cell organ, a "Golgi" body, while the other sees a coagulum resulting from poor staining techniques. In a similar example Hanson imagines Kepler and Tycho Brahe on a hill at dawn; Tycho sees the rising Sun but Kepler sees the rotation of the Earth.

Hanson's other examples are drawn from gestalt psychology. He mentions the Necker cube and a figure that may be seen either as a bird or an antelope (Fig. 3), similar to the duck-rabbit.[4] Hanson points out that we can be cued to see the figure in one way rather than another by putting it in a picture where other figures are more clearly birds (Fig. 4), or more clearly antelopes (Fig. 5), as desired. If it is right to describe all these "seeings" as observations, it cannot be right that observation will be the same for all suitably placed and equipped observers. The difference between the scientist and the student, or between Tycho and Kepler, is not a difference either in the orientation or scope of their visual field; nor is it a difference in

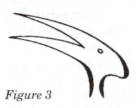

Figure 3

perceptual acuity. Rather, what makes the difference is a difference in knowledge or belief. If that is right, then observation is not prior to all judgment but is instead itself subsequent to, and dependent on, at least some judgments. The independence assumption, (iv), looks to be false.

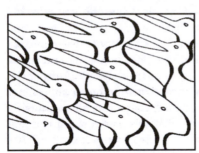

Figure 4

Figure 5

Hanson is concerned to resist a response that he regards as the natural one for the logical empiricist to make. That response denies that what is reported as "seeing an X-ray tube" really is an observation. Rather, so the response goes, this is a judgment subsequent to observation. The observer, in saying "I see an X-ray tube", is *interpreting* or inferring from what he or she observes. The basic perceptual experience, and observation, it is claimed, are independent of prior judgment and experience. The difference between the observers arises when, albeit unconsciously, there is an inference from this observation to a higher level judgment, and it is this inference that is influenced by the observer's background beliefs, experiences and so on. The logical empiricist may attempt to buttress this response by pointing out that in all the cases both members of the pair of (apparently differing) observers share the same pattern of retinal stimulation.

The latter point is quickly dealt with by pointing out that to have a retinal stimulation is not itself to see anything, or to make an observation; nor, in Hanson's words, is it a visual experience. (If drugged or distracted someone might have a certain pattern of retinal stimulation without observing anything at all, so retinal stimulation does not imply observation.)

The suggestion that the reports considered are interpretations of prior judgments Hanson seeks to rebut by reflecting on the nature of interpretation. Interpretation is, he says, a thought process. It typically takes time. It should be possible, in principle, to identify separately what it is that gets interpreted and the interpretation. But the cases of seeing are experiences, not the outcomes of thought processes; they are instantaneous, and the uninterpreted raw data are merely postulated, not empirically identified. Hanson regards the claimed existence of an (unidentified) uninterpreted experience that is instantaneously interpreted as a mere reiteration (without reasons) of the view "that the seeing of x *must* be the same for all observers looking at x".

Hanson's resistance to unconscious inferences is interesting. After all it is not altogether implausible that they take place. We can speed up our inferences by practice. For example an intermediate chess player will be able to make calculations about the consequences of certain moves more quickly than a beginner. The grandmaster makes such calculations in an instant. Perhaps part of what is going on is very rapid unconscious calculation? Let us imagine that this were true in the cases that Hanson presents us with, so the supposedly observational judgment is in fact inferred from more basic ones. The empiricist is saying that we are deceived by the rapid and unconscious nature of this inference into mistakenly thinking that the interpretation, the higher-level inferred judgment, is in fact the basic level observation. But if that were so, we would be violating assumption (iii), internalism about experiences. As we shall see shortly, this internalism is a feature of the empiricist program that both Hanson and Kuhn retain.

Thus the appeal to unconscious inference would be a poor strategy for defending the independence assumption. So putting this general defence of independence on one side, let us look at Hanson's positive arguments against independence. A significant element of Hanson's discussion, which plays a central role in these arguments, is a distinction between two senses of "seeing". In Hanson's usage two people can see incompatible things: Tycho sees a moving Sun and a stationary Earth, Kepler sees a moving Earth and a stationary Sun. The professor sees an X-ray tube, the schoolboy does not. One scientist sees a Golgi body, the other sees a coagulum. In ordinary usage we would not say this. Take the schoolboy looking at the unfamiliar piece of equipment. We would normally *not* say that he did not see the X-ray tube. He might point to or pick up the tube and say "What is this?" – which he could hardly have done if he didn't see it. Similarly, since the Earth is moving, Tycho did *not* see a stationary Earth, since there is no stationary Earth to see. Tycho *thought* he was looking at a stationary Earth. A visitor to New York might see and admire the Chrysler building without knowing that it is the Chrysler building. Indeed the visitor might even think that he is looking at the Empire State building. But in normal parlance we wouldn't say that he saw the Empire State building but not the Chrysler building; the very opposite, the visitor saw the Chrysler building and mistook it for the Empire State building, which was not seen.

Hanson recognizes that this is normal usage but claims that his sense in which the student does not see an X-ray tube while the scientist does is also a perfectly legitimate sense. There are two questions to ask: first, what is Hanson's sense, and how should we characterize it in contrast to the standard sense? Secondly, which of the two senses is the one relevant to the roles that "seeing" and hence "observing" play in science?

I shall call the normal sense "strong seeing", and Hanson's sense "weak seeing".[5] Hanson concedes that the schoolboy is "visually aware" of the X-ray tube, by which I understand him to be talking of seeing in the standard, strong sense. What then is visual awareness, strong seeing? Two things that the paragraph before last reveals about visual awareness are that: (i) one cannot be visually aware of what does not exist, and (ii) one can be said to be visually aware of X even though one does not possess the concept "X". I suggest that we should understand visual awareness of an object as a matter of being able to distinguish that object visually and to be able to discriminate it from other objects. This explains the two features just mentioned, in the first case because one cannot distinguish what does not exist, and secondly because one can discriminate what does exist from other things without having much in the way of conceptual apparatus to describe it. The visitor can tell the Chrysler building from its neighbours and so (strongly) sees it, even though the visitor doesn't have the concept "Chrysler building".[6]

Hanson's weak concept of seeing is different in both respects, but not illegitimate for all that. First, one *can* weakly see what does not exist – since Tycho can weakly see a stationary Earth, although there is no such thing;[7] one or other of the Golgi body and the coagulum does not exist, but both are weakly seen. Secondly, what seems to prevent the schoolboy from weakly seeing is the fact that the schoolboy is not able to place the X-ray tube under the concept "X-ray tube".

How does Hanson's weak sense of seeing X relate to the strong sense? It seems roughly equivalent to "believing that one is (strongly) seeing X". Note that it is *not* equivalent to "knowing that one is seeing X" since Hanson wants to allow it to be the case that Tycho sees a moving Sun and Kepler sees a stationary Sun. But since the Sun cannot be both moving and stationary, it cannot be that Tycho knows that he sees a moving Sun and Kepler knows he sees a stationary Sun. Nor do I think that "S weakly sees X" is equivalent to "S has a visual experience as of strongly seeing X". It is true that the latter would allow for weakly seeing what does not exist. I cannot strongly see a unicorn, but I could have a visual experience like that of seeing a unicorn. Nonetheless, this would not capture the conceptual dependence that Hanson wants for weak seeing. For I could hallucinate a unicorn or an X-ray tube, and so have a visual experience like that of strongly seeing them, but I need not have the concepts "unicorn" or "X-ray-tube" to do so.

Hanson states that weak seeing involves seeing *that*. He does not say exactly *how* they are related, but it is clear that part of the point of the connection is to introduce the conceptual element into weak seeing. "Seeing" and "seeing that" are different on any account; the former concerns objects ("seeing the Chrysler building", "seeing the X-ray tube"), while the latter concerns propositions ("seeing that the Chrysler building is tall", "seeing that the X-ray tube has two electrodes"). I'll call seeing that *propositional* seeing, and plain seeing, *objectual* seeing. One consequence of this difference between objectual and propositional seeing is that propositional seeing is *intensional* where strong objectual seeing is *extensional*. (For the difference between these see Note 8.) As mentioned I can strongly see an X-ray tube without having any *concept* of an X-ray tube, provided there is an X-ray tube there. However, if I see *that* the object is an X-ray tube, then I must have the concept "X-ray tube".[8]

The propositional character of "seeing that" is significant for epistemology. For if observations are to be the basis of scientific inference, then observations must be propositional in nature, since inferences take us from propositions to propositions. From the fact *that* the candelabra has a sharp corner, and from the fact *that* it is covered with blood and hair, the detective infers the conclusion *that* the candelabra is the likely murder weapon.

However, just as Hanson uses a weak sense of "seeing" he also uses a weak sense of "seeing that". In the strong sense of "seeing that" one can only see that *p* if it is true that *p*. Like "knowing that", "seeing that" is, we say, *factive*. If S sees that the liquid in the test-tube is boiling then it is true that the liquid is boiling. But if S is wrong, and the appearance of boiling is caused by bubbles from an impurity in the test tube, then S did not see that the liquid was boiling – he mistakenly believed he saw that the liquid is boiling. Hanson does not use this strong sense of "seeing that". His weak sense is non-factive: one can see that *p* even if it is false that *p*. Hence, according to Hanson, Tycho saw "that the Earth's brilliant satellite [i.e. the Sun] was beginning its diurnal circuit around us, while . . . Kepler and Galileo [saw] that the Earth was spinning them back into the light of our local star".[9] Since the two propositions that come after the two "that's" are incompatible, one at least must be false, hence, the "seeing" cannot be strong. Analogously with objectual "seeing", it seems that what Hanson means by (weak) "seeing that *p*" is something like "believing that one (strongly) sees that *p*".

So, to summarize, we may classify the different concepts of seeing so far discussed as being strong or weak, and as objectual or propositional. This gives us four possibilities:

	strong seeing	*weak seeing*
objectual seeing	S sees X, therefore X is an existing object	S sees X, but X may not exist (*)
propositional seeing (-that)	S sees that *p*, therefore *p* is true (*)	S sees that *p*, but *p* may be false (*)

The cases marked (*) are intensional cases of seeing, where the subject S must possess the concept "X" or the concepts contained in "*p*". Strong objectual seeing is extensional – one need not have the concept "X" to strongly see X. We also have an extensional cousin of strong propositional seeing. For instance one might truly say "John saw the Sun rise above Mt Cardigan" without John's having the concept "Mt Cardigan". Note that the word "that" is *not* used.

Seeing, empiricism and epistemology

Why does Hanson insist on these weak senses of seeing? Part of the answer is supplied by the fact that they offer what looks like a quick and easy route to rejecting the independence assumption. The strong sense of objectual seeing allowed that similarly situated observers observe the same thing, even if they employ different concepts to describe it. To defeat independence Hanson wants to argue that observers see different things depending on their background beliefs. One way background beliefs come into play is by providing a conceptual apparatus. Thus Hanson wants to use an intensional sense of seeing so that a difference in concepts yields a difference in what is seen.

I think, nonetheless, that Hanson has a deeper reason than this for preferring the weak senses. The quick route to defeating independence depends only on the intensionality of weak seeing. This rules out using strong objectual seeing, since that is extensional. Note, however, that strong propositional seeing is also intensional, and so the same result could have been obtained without using a weak sense of seeing. Two observers might propositionally see different things if they have different concepts, even in the strong sense of "see".

So a preference for intensional seeing is not sufficient to explain a preference for the weak senses of seeing. The important thing about the weak senses is that they are neutral with respect to truth, in the case of propositional seeing, and existence, in the case of objectual seeing. Hanson accepts the empiricist premises (ii) and (iii). A report of what one sees is an observation report, hence, given premise (ii), *seeing is an experience*. Premise (iii) says that the nature and content of experience must be internally accessible – it must be possible for subjects to know what experiences they are having. If that is so, *experience cannot depend on something independent of the subject*. Strongly seeing that the liquid is boiling can take place only if the liquid is boiling. But since the boiling of the liquid is something that takes place outside the subject, it cannot, according to this view, be part of the subject's experience. Hence strong propositional seeing cannot be an experience.

It is highly significant that Hanson, while seeking to undermine the empiricist assumption of independence, is nonetheless committed to the empiricist premises (ii) and (iii), in particular the latter, internalism about experience. It is this that leads him to take observation to be neutral with regard to truth and existence. Kuhn, we shall see, inherits all these assumptions. Correspondingly, Kuhn takes a neutralist view also, which is one reason why he refuses to regard truth as relevant to the explanation of scientific change. We shall see in Chapter 6 that he has a notion of scientific progress, but this is nothing to do with increasing truth or nearness to the truth. We can usefully sketch the pattern of Kuhn's thought here: as described in the last chapter, the driving forces behind scientific activity are perceptions of similarity or dissimilarity between exemplars and proposed puzzle-solutions. Since (weak) perceptions are neutral with regard to truth, truth is irrelevant to the operation of the force behind scientific change. Hence there is no reason to suppose that scientific change drives us towards truth. This same neutralism and a truth-independent notion of progress invites scepticism or relativism.

The supposed importance of an experience being internal is that if it is internal then, it is claimed, one cannot be mistaken about whether one has it or not.[10] Their infallibility is why empiricists were happy to take reports of experiences to be the basis of scientific inference – premises (i) and (ii). It has been traditional therefore to break strong seeing (and knowing) down into at least two components: an internal, experiential component that is neutral regarding truth, and the external truth of the propositional content (i.e. a fact). (Some further component may also be required, such as a causal connection or justification.) Thus in the case of knowledge, the experiential component is belief, so knowledge is some kind of belief which is true. Similarly Hanson implicitly regards strong propositional seeing as having an experiential component, viz. weak propositional seeing, plus (at least) truth.[11]

As for the experiential component, weak seeing, Hanson regards this as having two aspects: one is the pictorial or "optical" and the other is linguistic. The thrust of Hanson's argument has been to contend that what is *not* happening in vision is that one's basic experience is just the pictorial component, with the linguistic/conceptual element coming later as a result of interpretation. According to Hanson one cannot have the one without the other. Seeing is inevitably both together. Thus while Hanson thinks that the truth element of strong seeing can be peeled off to leave just weak seeing, he denies that the conceptual element of weak seeing can be removed to leave pure unconceptualized optical experience. Intensionality is built into experience.

What may we conclude about observation? Hanson wants us to believe that observation is theory-laden in the sense that it comes unavoidably conceptualized, and furthermore the concepts employed are ones that very often will derive from our theoretical beliefs. Hence the nature of observation will not always be common to all suitably placed observers. This depends only on the inherent intensionality of experience, a view that can be granted to Hanson whether one is thinking about experience and observation in a strong or weak way. To establish this Hanson does not need also to adopt internalism and so neutralism about experience. The fact that he does so is a sign of significant residual empiricism.

It is worth pausing to reflect whether retention of this facet of empiricism is an advantage to epistemology. On the one hand, internalism is supposed to provide epistemology with secure foundations, since one supposedly knows what one's experiences are. Even though the logical empiricists and Hanson all took it to be true, this view can and ought to be challenged.[12] Whatever its alleged advantages, this sort of foundationalism suffers from long recognized difficulties. Even if we accept internalism, this gives subjects justification only for reports of their experiences. Any attempt to found knowledge on such reports faces the difficulty of justifying any inference from them to the sorts of proposition that science is interested in, which typically concern objects and processes rather far removed from the subject's psyche. The same point can be made viewed from another angle. Imagine a witness in a criminal trial who claims to have seen the defendant near the scene of the crime, while in fact the defendant was elsewhere. Hanson may say that the witness weakly saw the defendant at the crime scene, but we must deny that he strongly saw this, since one cannot strongly see what is false. Clearly we would not want to start inferences from the observation that p where that observation is a report of the experience of weakly seeing that p, since we may thereby start from falsity. If we want our inferences to start from truth, we should prefer them to start from observations of what we strongly see.[13]

Empiricism and its foundationalist cousins that seek to base knowledge on internalist experience are always hostage to the danger that their position will collapse into scepticism. It is not surprising that so many empiricists were either sceptics, or sought to avoid scepticism by watering down reality of things in order to make them more readily known, as is characteristic of idealism. Idealism can be regarded as a kind of relativism, since it holds that what exists and what is true depends on the subject's states of belief or experience. Since it is the internalism of empiricism that is responsible for this, an internalism shared by Kuhn, it is not surprising that Kuhn is at the very best on the edge of scepticism and relativism. Sometimes Kuhn seems to be avoiding saying anything on the epistemological issues of what we know and whether our beliefs are getting closer to the truth, or on related metaphysical ones of whether what there is is independent of our experience – an approach that fits with a strict application of neutralism. However, when he breaks his silence on such issues he tends, as we would expect given his internalism, to cross the line into either scepticism or relativism or both.

Kuhn on perception

In the last section I sought to show how Hanson aimed to undermine one important assumption of logical empiricism but at the same time continued to adhere to others. I also claimed that the same is true of Kuhn. In this section I shall make good this claim. Kuhn draws heavily upon Hanson. Not only does he cite Hanson,[14] but he employs Hanson's examples and related ones intended to make similar points. In addition to the gestalt examples of the Necker cube and the duck-rabbit, Kuhn discusses a physicist who sees a record of familiar sub-nuclear events when looking at a bubble-chamber photograph, while the student sees only confused and broken lines. This case, and that of a cartographer who sees a picture of a terrain where the student sees only lines on paper, remind us of Hanson's difference between the scientist and schoolboy in their perceivings of the X-ray tube.[15] Kuhn also has cases analogous to Hanson's Tycho-Kepler disagreement: "Lavoisier . . . saw oxygen where Priestley had seen dephlogisticated air and others had seen nothing at all".[16] "Where Berthollet saw a compound that could vary in proportion, Proust saw only a physical mixture."[17] Kuhn devotes an extended discussion to a "shift of vision" between the Aristotelian and Galileo, when looking at swinging bodies: "To the Aristotelians, who believed that a heavy body is moved by its own nature from a higher position to a state of natural rest at a lower one, the swinging body was simply falling with difficulty . . . Galileo, on the other hand, looking at the swinging body, saw a pendulum, a body that almost succeeded in repeating the same motion over and over again ad infinitum".[18]

Kuhn also draws upon psychological discoveries not mentioned by Hanson. In particular he discusses a set of experiments carried out by Bruner and Postman.[19] Their fundamental claim is that perception is a process that occurs upon the stimulation of a "prepared" subject. The subject is prepared in that he or she comes with prior expectations, needs and so on. In particular, the processes that lead from stimulation to perception, processes directed by the preparedness of the subject, tend to organize experience so that expectations are met and features relevant to current needs are prevalent, while experiences that go against expectation or are irrelevant to need are minimized. The researchers showed playing cards to subjects for varying, but short periods of time (from 10 to 1000 milliseconds). The subjects were asked to record all they saw. Some of the cards were normal, but some were incongruous. For example, one card had the form of a six of clubs, but was coloured red not black; a four of hearts was not red but black instead. Their central finding was that the time required to give a correct description of the cards was much higher for the abnormal cards than for the normal ones – on average 114 as opposed to 28 milliseconds. Furthermore, the time required to correctly identify an abnormal card fell if the subject has seen an abnormal card before (and it rose slightly if the card followed a sequence of normal cards). At shorter exposures especially subjects failed to make correct identifications. Bruner and Postman characterized three kinds of failure. The most common they called *dominance* – the subjects would give the description of a normal card. For example, given a red six of spades, a subject might say "six of hearts". Here subjects have noticed the colour, and so called a suit that fits with that colour according to their expectations. Or they might say "six of spades", but not notice the unusual colour, allowing the perception of shape to dominate. In *compromise* reactions, the subject reports a "perceptual middle ground" somewhere between what the card actually is and what the subject expects. So the colour of an incongruous card might be reported as "brown", "purple" or "greyish red" and so on. The third failure of correct identification is *disruption*, which may occur at exposures longer than those for which a confident but erroneous dominance or compromise reaction is given. The subject may become confused and be unable to give a clear report of his perceptual experience: "I'll be damned if I know now whether it is red or what!"

It is clear that what subjects perceive, or think they perceive, is in part determined by prior experience. If we permit ourselves a weak

109

sense of perception, then perception is in part experience- and expectation-dependent. More generally we should think of the (weak) perceptions of subjects – let us say their *perceptual experiences* – as being fixed by two factors: the stimulation and their existing psychological state. Kuhn however draws further conclusions, beyond those demonstrated by Bruner and Postman. He sees the Bruner and Postman experiments as analogous to, or even illustrative of, the same kind of mental process as is involved in a shift of scientific paradigm. Ultimately he extends shifts in perception to encompass changes of "world". Referring to the student who has now learnt how to understand a bubble-chamber photograph, Kuhn says: "The world that the student then enters is not, however, fixed once and for all by the nature of the environment, on the one hand, and of science, on the other. Rather it is determined jointly by the environment and the particular normal-scientific tradition that the student has been trained to pursue".[20]

The leap to this from the Bruner and Postman experiments is large and merits comment. An intermediate step is provided by the Lavoisier-Priestley oxygen and the Galileo-Aristotle pendulum cases. Kuhn describes these cases, as Hanson describes the sunrise case, as involving differences in perception. The differences in experience among the Bruner–Postman subjects can be explained in terms of their prior perceptual experience. It is because they are accustomed to *seeing* only black spades, that they find it difficult to experience a red spade as such. In ways that we have discussed already, concerning learned similarity and pattern recognition, a *visual* expectation is created by exposure to *visual* exemplars. But in Kuhn's cases the alleged difference in vision does *not* originate in facts about prior visual experience. Rather, the source is a difference in theoretical beliefs. Thus Hanson describes seeing in such cases as "theory-laden": "Observation of *x* is shaped by prior *knowledge* of *x*". Note that Hanson says "knowledge" not "experience". Thus we are well on the way to thinking that perception is paradigm-dependent. And given a suitable definition or conception of "world" (as a quasi-technical term) in terms, in part, of experiences, then we have the conclusion that the world is paradigm-dependent.

It is now worth looking at the examples in detail, to see to what extent they support such conclusions.

Gestalt images

Hanson's use of gestalt images helps show that the ways things seem to us is not as raw sense data but often as a product of our conceptualization of things. Our basic sense experience is not an experience awaiting order to be put upon it but comes ready ordered. Hence the duck-rabbit appears to us as a picture of a rabbit or as that of a duck. Rarely does it appear as a mere curly line and a dot. In particular, certain beliefs or cues can make us see the picture in one way rather than in the other. Kuhn himself warns that the gestalt cases may be misleading. He points out that we can learn to see the duck-rabbit just as a lines and a dot.[21] In other cases, where there are misleading cues, we can learn (or be told) what it is we are really looking at. We can then say, "I really see the lines, but I can see them alternately *as* a duck or *as* a rabbit". In effect, we can learn to see things in a way that is akin to the uninterpreted (or, better, less interpreted) way the logical empiricist thinks we actually do see things. The very point of these psychological demonstrations is that they allow analysis in this way. What is going on is that there is an external standard that allows us to investigate what is being looked at in a neutral way, one that is independent of seeing the image as a rabbit or duck. By contrast, says Kuhn, in the scientific case we are not in a position to do this. We do not have some independent source of information on what we are looking at that can tell us that the way things seem to us to be fails to reflect the way they really are, or reflects a mental imposition on the objective facts. Furthermore, the scientist cannot flip back and forth as one can between the two images. For this reason, a scientist may not properly be aware of a perceptual shift that accompanies a paradigm shift.

Kuhn's point is that if we take the gestalt analogy with scientific belief too earnestly, we may end up oversimplifying the scientific cases, and may even return to something closer to the empiricist view. For we might suppose that we might be able to get to a more basic neutral view of paradigms, akin to seeing the duck-rabbit just as a line and a dot. But Kuhn denies that this is possible – it is as if we must see the duck-rabbit either as a duck or as a rabbit, but never as neither, nor as both. While Kuhn wants to warn against the dangers of reading the duck-rabbit case in a way too sympathetic to logical empiricism, it is true also that the discussion of gestalt images may be misleading in the opposite direction. Since we are dealing with drawings and images there is no question of a correct or incorrect perceptual belief. The duck-rabbit picture is neither a duck nor a rabbit and no one takes it to be. For that matter, there is no right answer to the question "Is the duck-rabbit picture a picture of a duck, or of a rabbit?" in the way that there is a right answer to the question "Of whom is the picture known as 'La Giaconda' a portrait?" Hence it is perfectly appropriate to use the weak sense of "see" in describing people's experiences on seeing these images; since there is no room for issues of right and wrong, there is no room for the stronger sense. It is natural to say that a subject at some moment sees the image *as* a rabbit. Hanson says that he doesn't want to identify seeing and seeing as. Nonetheless he thinks that seeing as is part of the logic of seeing, and says that the logic of seeing as illustrates the general case of seeing.[22] It is however highly contentious to think that the general case of seeing is like one where issues of truth and falsity are transparently irrelevant. The latter would in turn suggest that the essential nature of perception is exhausted by the way things appear to be to the subject. It seems again that a commitment to internalism underpins what Kuhn and Hanson say. Putting the issue of internalism on one side, if one is not yet decided on whether truth and falsity are irrelevant to perception, then one ought to be wary of taking too much from the gestalt cases.

Student learning cases

I have already given some discussion to Hanson's case of a schoolboy coming to recognize an X-ray tube as such. One conclusion readily drawn from that discussion is that where intensional readings of "see" are involved, what one can "see" will depend on what concepts one has. It may be that what concepts one has available in turn depends on which theories one is acquainted with and so this seeing may thus be theory-dependent. This case is not identical to the gestalt ones. As Hanson admits, even the ignorant schoolboy is able to see that various things are true of the X-ray tube, for example that it will break if dropped, and so it is clear that he sees it as a thing, a unity, as having a form or gestalt. Let the schoolboy learn the meaning of "X-ray tube", perhaps by a description of its functioning, so that he has a concept of an X-ray tube. Now he has the capacity for seeing the X-ray tube in intensional respects too. But it does not appear that this acquisition of a new concept changes the subjective perceptual experience he has in any significant regard.

Kuhn's cases are different in a very important respect, for it is not merely a lack of the expert's vocabulary that is doing the work. We may imagine that student geographers are able to recognize different geographical features when they encounter them in the field, and apply the appropriate concepts to them; it is just the cartographic representation that they cannot yet understand. Similarly physics students may know a lot about electrons, protons and so on but have not yet seen a bubble-chamber photograph. In these cases students need to learn to read the map or photograph; after some experience they will become as proficient as the experts.

It is worth noting what goes on in such learning processes. Students, once told how the contour lines function, will be able to work out that such a configuration is a valley, another an escarpment, that this point is visible from another, and so on. It is inevitable that this starts out as a matter of mental calculation and inference. With time the inference gets quicker and eventually the reading of the map is immediate and intuitive. At the end of this learning process it is correct to say that they now see that this is a saddle and that a ridge. And to this extent, unlike the X-ray tube case, these cases have something in common with the gestalt examples. Students see a form where before they saw none. Their experience is importantly different. Note though that these cases make the contrast with empiricist claims about interpretation less sharp. For in the early stages there is obvious interpretation taking place. Consequently it looks as if we can say of the end of the learning process *either* that the interpretation has disappeared, it being no longer necessary, *or* that it has become speedy and unconscious. I am inclined to side with Kuhn here – even if there is fast unconscious interpretation of raw data, the grounds are thin for regarding such data as constituting the subject's experience of the things discussed.

Such cases, along with our tendencies with respect to playing cards, are best regarded as instances of recognition through habituation – acquiring a recognitional capacity or a similarity relation through training with exemplars. While this is a matter of learning to see a form or gestalt, the source of our experience of the Necker cube is not easily understood in precisely this way, since we do not learn to see a cube, nor is experience an explanation of why we can switch between the two gestalts of the cube. The oxygen and pendulum cases are, as we shall see, rather more contentious. Our expectation, that a card (which is a heart) is red, is based primarily on the fact that all our experiences of playing cards have reinforced this expectation. It is best to think of the generalization "all hearts are red" as hard-wired in such a way that the nature of experience is directly affected.

113

Why is Hanson's X-ray tube case different from Kuhn's learning cases? The relevant difference between the schoolboy and the professor is simply that the latter has a concept "X-ray tube" that the former does not. Hanson has engineered his notion of weak seeing so that it is sensitive to changes just in conceptual apparatus. We might be inclined to think that there is some important notion, indeed the natural notion of visual experience that is not so sensitive to simple conceptual change, which may be captured when we say: being able to call the instrument an X-ray tube doesn't change the way it looks to the boy. Above I noted that there is a weak sense of seeing, or sense of visual experience, which can be roughly characterized as: "having an experience as of strongly seeing X" of "having an experience like that of strongly seeing X" which can be had without having the concept "X". This may very well be the same for the professor and the schoolboy. But it is rather less clear that the visual experience, in this sense, remains the same for students learning to read the bubble-chamber photograph, for after training students are able to see, to have visual experiences of, certain patterns which they were not able to see beforehand.[23]

Phenomena and theory

The most contentious cases are those concerning rival scientists' perceptual experiences of various phenomena: Hanson's Tycho–Kepler sunrise case and Kuhn's two cases of Galileo and Aristotle on the pendulum and Lavoisier and Priestley on oxygen. Before proceeding too far we should consider the possibility that Kuhn is here using perceptual verbs merely metaphorically: there is no literal difference in visual experience between Priestley and Lavoisier when they look at a jar of oxygen. It should be emphasized that Kuhn's text gives every indication of being meant literally.[24] He not only says that Lavoisier saw oxygen where Priestley had seen dephlogisticated air but prefaces this by saying that it is a "transformation of vision" while he describes the difference between the Aristotelian and Galileo as a "shift in vision".[25] Kuhn recognizes that not everyone is likely to accept these claims, asking, "Do we, however, really need to describe what separates Galileo from Aristotle, or Lavoisier from Priestley, as a transformation of vision? Did these men really *see* different things when *looking at* the same sorts of objects?"[26] He admits that "Many readers will surely want to respond that what changes with a paradigm is only the scientist's interpretation of observations that themselves are fixed once and for all by the nature of the environment and of the perceptual apparatus". This is exactly the response that, as we have seen, Hanson considers in his discussion of perception and which he rejects; Kuhn also rejects this view, as being part of a Cartesian paradigm in philosophy that has been shown to be askew. He thereby implies that a better understanding of perception will show us that these locutions, unfamiliar though they may be, are entirely appropriate and not metaphorical.

Even so, if Kuhn's usage were metaphorical then certain remarks would be pertinent. First, we would be entitled to ask, what work is this metaphor doing? If Kuhn's use of "see" signals simply a difference in what Priestley and Lavoisier think the stuff in the jar is, and so a difference in how it fits into their theoretical and experimental beliefs and practices, then his claim that they "see" different things is uncontroversial and could be accepted happily by a positivist. To be interesting, there would need to be a way of articulating the metaphor of "seeing differently", if it is a metaphor, that is weaker than literally seeing differently but stronger than just believing differently. Kuhn does not supply us with such an articulation, though towards the end of the chapter I suggest a possible answer. Secondly, if Kuhn accepts that there is no literal difference in observation of the stuff in the jar then we have no reason to suppose that the theory-ladenness or theory-dependence of observation is an issue in this case. Remember that one of the roles that the independence assumption is supposed to play in the empiricist account is that of ensuring that observation can be the final court in choosing between theories. Theory-dependence undermines independence and hence the ability of observation always to be the arbiter of theory choice. But if observation in the chemical laboratory is not dependent on whether one belongs to Priestley's or to Lavoisier's paradigm, the possibility of arbitration by observations common to both chemists is not ruled out.

It is important therefore that we see whether there is any way in which visual experience can be literally different between possessors of different theories. Earlier in this chapter I distinguished between the X-ray tube case where the *mere* acquisition of a new concept did not make an appreciable difference to the content of experience, and the learned recognition of geographical phenomena in maps or subatomic ones in bubble-chamber photographs. This distinction will help in unravelling what is going on in the present set of cases, where, according to Kuhn and Hanson, scientists with rival theories or paradigms see different things. The distinction allows us to see that mere possession of different vocabularies and conceptual apparatuses is not itself sufficient to produce a difference in visual experience, as naturally understood, even though it does produce a difference in what the subject can intensionally "see" and, thus, in what the subject reports himself or herself as seeing.

Should the three cases of the sunrise, oxygen and the pendulum be regarded as *mere* conceptualizations, like the X-ray tube case? Or as involving genuine shifts or differences in perceptual experience? Inspection of the cases should incline us strongly to the former. Speaking for myself I have no sense that knowledge of the Earth's rotation affects the visual experience I have at dawn. If anything it still looks to me as if the Sun is moving as Tycho and Ptolemy believed to be actually the case. Not all relevant knowledge has an effect on perceptual experience, even when that knowledge tells us that the experience is misleading. The horizontal parts of the famous Müller-Lyer lines (Fig. 6) still look to be of different lengths even when we measure them and know they are the same. The Sun still looks to be moving even when we know it is not. Imagine a person for whom the Sun is obscured by the brow of a hill and who walks up the hill until it is visible. If Hanson were right in thinking that Tycho and Kepler have different experiences, the walker's experience should be like Kepler's (since both know that the Sun comes into view because of their own, not the Sun's movement) and so should be unlike Tycho's experience. But this seems implausible – the two most similar experiences are Tycho's and Kepler's, while the walker's experience, to the extent that it is different is different equally from both. It is true that Kepler may think of what is going on as being similar to what is occurring to the walker. But that thinking is not part of the experience. Kepler employs what we called a "similarity relation", but not because it is a relation learnt by repeated exposure to similarities of this kind, but instead because of Kepler's theoretical beliefs.[27]

Kuhn's case of Priestley seeing dephlogisticated air where Lavoisier saw oxygen is even less plausible as an instance of difference in perceptual experience. A glass jar filled with nitrogen is

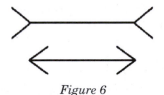

Figure 6

perceptually indistinguishable from one filled with oxygen, even if one knows which is which. Similarly there is no reason to think that there is a perceptual difference when two people look at the *same* gas but give it different names. The same remarks are pertinent to Kuhn's claim "Where Berthollet saw a compound that could vary in proportion, Proust saw only a physical mixture".[28] There is no difference in visual experience here, beyond the use of different concepts. It is like the X-ray tube case, but unlike the bubble-chamber and map cases, or the playing card case.

For the most part, similar comments are called for in the case of the apparatus that Galileo regards as a pendulum and which the Aristotelian takes to be an object in constrained fall. *Why* should things look different to them? It is perhaps possible to stretch this case to near coincidence with Kuhn's verdict. For it is plausible to say that causal relations can be part of our experience of things. In seeing a racket hit a ball we do not see just a motion of the racket and a change in the motion of the ball, but we see the one strike the other; we see the change of the motion of the ball as being the result of the striking by the racket. Even though we do not normally see air we can plausibly say that in seeing a leaf being blown about by the wind, the fact of its being moved by the wind is part of the experience of seeing the leaf. Indeed we will often say "look at that wind" pointing to bowed trees and swirling leaves. It may not be impossible therefore to think that Galileo and the Aristotelian have different experiences since they think of the motion of the pendulum bob as being caused in different ways.

Even so, this is only a somewhat strained story about how Kuhn *might* be right. It certainly does not show that he is, for not all causes form part of our experience of their effects; nor do our beliefs about causes always influence our experience of events. Furthermore, even if there is *some* difference in experience, it needs to be shown that it is philosophically significant. It is relevant to the task of debunking positivism and logical empiricism more generally. But it has also been regarded as an important step on the road to demonstrating some important problem of incommensurability, even relativism. These topics shall be addressed in the next chapter. For the time being it should be noted that even Galileo and the Aristotelian may be able to agree on many things, such as the average period of a set of swings.

There is little reason to think that in *these* cases one's theoretical beliefs or knowledge affect one's perceptual experiences. There is therefore no general rule that perceptual experience is theory-laden or paradigm-dependent. And if there is no general rule that observation is dependent, then it is possible that on some occasions at least observations will be able to decide between competing theories or be an appropriate measure of the quality of both an older and a newer paradigm.

Nonetheless, it is true that the competing scientists may choose to describe what they see using different vocabulary, and may thus make (what they take to be) observational statements that are incompatible. This does not itself show any difference in experience. Thus when Kuhn says "A sees x" while "B sees y", we should not regard this as reporting a difference in experience between A and B. Instead a weak, intensional, reading of "sees" is being used, one equivalent to "believes he/she (strongly) sees". But then there is nothing at all contentious about saying that Priestley believed he saw dephlogisticated air while Lavoisier believed he saw oxygen.

Summarizing the conclusions of this section, Hanson and Kuhn have undoubtedly shown that in a normal sense of "visual (or perceptual) experience" what a subject can experience is dependent on their prior experience. The Bruner and Postman experiments establish this admirably, and Kuhn's bubble-chamber and map cases illustrate it convincingly too. The empiricist's independence assumption (iv) is false. However, this sense of visual experience is *not* the same as Hanson's weak seeing. The latter is intensional and thus is trivially sensitive to change in the subject's concepts. To show that in this weak sense two people see different things is not to show that they have a different visual experience in the natural sense.

There is therefore an equivocation in Kuhn's discussion, when we move from talking of a visual shift in the case of learning to read a bubble-chamber photograph to talking about a visual shift in the transition from being an Aristotelian to a Galilean, or from Priestley to Lavoisier. These are different shifts, one a shift in experience, naturally understood, another a shift in concepts employed. "What a man sees," writes Kuhn, "depends on what he looks at and also upon what his previous visual-conceptual experience has taught him to see."[29] My argument has been that the dependence is quite different, and deeper, when the prior experience is visual, from what it is when the experience is merely conceptual. Talking of "visual-conceptual experience" blurs this difference.

Earlier I suggested that the notion of strong seeing is tied to abilities in making visual discriminations, detecting things and so on. I further suggest that the natural notion of visual experience is similarly tied to these abilities, albeit more loosely. This is why the bubble-chamber and map cases are instances of change in visual experience – the students, after training, can make visual discriminations, such as spotting certain salient patterns, that they could not before. Mere conceptual shift need not provide this. It ought nonetheless be noted that sometimes these two go hand in hand. When we acquire concepts by ostensive definition, learning concepts with samples, we often acquire a recognitional (and hence discriminatory) capacity at the same time. Learning the concept "sheep" will involve seeing sheep and pictures of them in such a way that we can tell them apart from dogs, pigs, goats and such like.[30] So some instances of conceptual change will bring with them changes in experience. As we saw in the last chapter, Kuhn thinks that acquisition of a paradigm is a case of learning with samples (exemplars). It is perhaps natural then to think that acquisition of a paradigm will bring both conceptual and perceptual change bound inextricably together. But that would be a fallacy; even if a paradigm is acquired that way, it would bring perceptual change only if the learning with exemplars resulted in a change in *visual* discriminatory powers. That might be achieved if the exemplars had a strong visual component, which was used in the acquisition of concepts by ostensive training. That, undoubtedly, occurs for some scientific concepts. But for the vast bulk of theoretical concepts it does not. No visual training was involved in my learning of the concepts "neutrino", "gene", "supernova", "isotope", "enantiomorph", "molybdenum" and so on.

Theoretical influences on perception

Kuhn and Hanson have shown that past visual experience and learning can affect the nature of current visual experience. They have also shown that changes in theory and hence in concepts can change the way we conceptualize our experiences. My claim has been that these two true theses are distinct but that Kuhn and Hanson have tended to elide them, in large part by uncritical characterization of experience in terms of weak seeing. This elision leads to the conclusion that a change in a theory can lead to a substantial change in a perceptual experience that is more than a change in the concepts used to express it. Elision is no substitute for argument, and no argument has been presented to close the gap.

Nonetheless, it may still be the case that there is a link, that theoretical beliefs can affect visual experience. For example, at the end of the last section I assumed that a visual discriminatory capacity could only be acquired by training with visual examples. Could not some such capacity be acquired by non-visual, theoretical learning? This is an interesting question, and the possibility cannot be excluded *a priori*. We could carry out a version of the Bruner and Postman experiments on subjects who had no prior experience of playing cards. We first train them (visually) to recognize the various suit symbols, using just outlines of the shapes, so that this acquisition gives the subjects no association of colour and suit. We then inform the subjects in some non-visual way that playing cards are such that hearts and diamonds are always red, spades and clubs always black. This is a little bit of theory. Finally we carry out the tests as Bruner and Postman did. Would we find the same sorts of response to the anomalous cards that they found in their subjects? If we found some difference in their ability to describe the normal and the abnormal cards, then we might conclude that theoretical belief can influence the nature of experience.

There is indeed evidence of this kind, although not from the imagined experiment just described. It is reported, for example, that people who believe that muggers are more often black than white are more likely than those who do not have this belief to judge mistakenly that the man with the knife in a mugging was black. Scientific cases show something similar. The case of Blondlot's N-rays is one of the most notorious. In 1903 René Blondlot claimed to have discovered a new kind of ray, instances of which were recorded and investigated by a large number of eminent French scientists. Outside France interest in N-rays waned when it was reported by the American physicist R. W. Wood that during a visit to Blondlot's laboratory he surreptitiously removed from the apparatus an essential prism. Despite the secret sabotage of his equipment, Blondlot still reported seeing the effects of the N-rays. It seems plausible to suggest that Blondlot's visual experiences were different from, for example, Wood's, and that this was related to his theoretical belief in the existence of N-rays. In a study conducted by Robert Rosenthal, students were asked to carry out experiments on two groups of rats, a "maze-bright" group and a "maze-dull" group, which they were told differed genetically in their ability to navigate mazes. Sure enough the students found significant differences in the times the rats took to find their way about mazes. In fact Rosenthal had selected all the rats from one and the same strain. The difference in the students' results originated not in the rats but in their beliefs about the rats.[31]

Kuhn does not himself give us psychological evidence of this sort. Can we build a case for Kuhn from it? I do not think so. First, the effects in question would have to be strong and fairly pervasive to make a difference to our discussion. Kuhn wants changes in a paradigm to lead to systematic changes in perceptual experience. My suspicion is that in the imaginary version of the Bruner and Postman experiments the effect of theoretical learning would be much weaker than the effect of visual learning. As regards the original experiments, the maximum period for exposure was one second – most exposures being rather shorter than this. The effect, though real and important, is marginal. It is easily swamped by extended exposure. Similarly in Rosenthal's study it required only slight mis-recording to get the results the students obtained. This too could have been swamped by a correspondingly greater real difference between the rats. A prejudice-driven mistake about a person's colour is more likely to occur when there is some cause for doubt (the glimpse was fleeting or the lighting poor) than when circumstances are optimal for perception. So although such cases may well be common it is implausible to see them as an essential feature of perceptual experience.

Secondly, it is not entirely clear in all such cases that there is a difference in perception as opposed to a difference in what is inferred or what is done. For example, did Blondlot really have a different perceptual experience? Or did he select things he was in fact was experiencing as being N-rays or caused by them? In Rosenthal's study it seems that much of the effect came not so much from (mis-)-perception but from unconsciously biased time-keeping – pressing the stop-watch fractionally too late for the supposedly dull rats, too early for the bright ones.

Thirdly, the effect in some of these cases is best understood not as deriving from some especially close relationship between belief/ theory and perceptual experience but instead as exemplifying a more general phenomenon: wishful thinking and self-deception. We often have unfounded non-perceptual beliefs because we want them to be true, and this extends to perceptual beliefs too, especially in percep-tually marginal or difficult cases. Blondlot's beliefs as well as those of his French colleagues are best explained by a national pride, sensi-tive after the discoveries of Roentgen's (German) X-rays and Crookes' (British) cathode rays, that was desirous of French rays too. I suspect that mistakes about a mugger's skin colour will be more prevalent among those with racial prejudices than those without, even when statistical beliefs are shared. In double-blind trials of new drugs and other medical procedures the researchers do not know which patients are receiving the drug under test and which are receiving the placebo. Biased results may occur without the double-blind but this is probably due less to an expectation of the efficacy of the drug than to a desire for a successful outcome to the trial. Even in Rosenthal's study self-deception is not far away, since students want to get results that accord with what is expected. At some level in our psychologies the expected is more comfortable than the unexpected, and so to that extent is desired and may potentially be affected by self-deception.

All in all, the case for an important link between theoretical belief and perceptual experience, as opposed to a marginal link or one that is mediated by self-deception, is weak. What may we conclude concerning Kuhn's attack on the empiricist understanding of observation? Even the existence of a weak influence of theory on perception, along with the stronger effect of visual learning that Bruner and Postman have identified, is enough to show that the independence assumption is strictly false. It is certainly no *a priori* truth. The empiricist view is badly undermined. Kuhn's purpose in this attack is to show that paradigm change cannot necessarily be arbitrated by the neutral court of observation, for observation need not be neutral. This weak claim should be granted. Has Kuhn established the stronger claim that in fact observation is not neutral between paradigms, that there are substantive disagreements in perception that prevent shared observation? This has not been shown as a general claim about paradigm shifts. Even if true as regards paradigms involving a considerable amount of visual training, we have seen that it is not true as regards paradigms whose content is primarily theoretical. The effect of theoretical belief on perceptual experience is weak and marginal. It is not strong enough to affect perception in a way that prevents observation from being common to competing paradigms (such as those of Priestley and Lavoisier, or of Newton and Einstein), and so preventing it from playing a part in deciding between the paradigms. There is no reason to think that the perceptual experiences of Eddington and his colleagues, who observed the solar eclipse of 1919, were affected either by their training in the Newtonian paradigm or their commitment to the Einsteinian paradigm which their observations supported.[32]

Worlds and world changes

One of the most famous but least well understood theses in *The Structure of Scientific Revolutions* is that the world changes when paradigms change. The latter idea makes several appearances in Kuhn's work, but he never gave a detailed account of it. On one occasion Kuhn states the thesis with a disclaimer: "In a sense that I am unable to explicate further, the proponents of competing paradigms practice their trades in different worlds".[33] Kuhn introduces the world change thesis as being closely linked to his views on perception and observation, while it is also clear that he also regards paradigm-induced world changes as aspects of the general phenomenon of incommensurability.[34] In *Structure* the view is that in some more or less literal way the world looks different after a paradigm change, and so the way the world looks (which gives us our observations) cannot be the measure of theories – they are thus incommensurable. In his later work Kuhn's account of incommensurability becomes more clearly linguistic and less perceptual; I address the linguistic account in the next chapter. That shift itself illustrates the elision I have mentioned before between the perceptual and the conceptual in Kuhn's thinking. That close link is characteristic of empiricism, which tends to see meaning as rooted in perceptible things and properties. It may be that the final dropping of the perceptual element in the account of incommensurability shows some recognition of the difficulties. The world change thesis is correspondingly less prominent. In this section and the next I shall see whether a perception-based account of world change is viable. I conclude that there is not much room for a significant thesis of this kind. Nonetheless I suspect that Kuhn was onto something, even if, as he says in the admission quoted, he was never able to satisfactorily explicate it. In a later section I make a suggestion as to what he might have been aiming at. World changes are neither concerned exclusively with perception nor are they linguistic; instead they relate to more general quasi-intuitive properties of the mind. These are readily understood on the sort of connectionist model of the mind that I appealed to in explicating Kuhn's views on the functioning of paradigms.

At first glance the world change thesis seems to imply some form of idealism, the view that there is no world beyond what we see or believe in. One version of idealism that has been (mis-)perceived in Kuhn says that the world is just what the current paradigm says it is. This view, a strong version of what is called social-constructivism, is not Kuhn's. If the world is *merely* what our paradigm says it is, it becomes difficult to see how we make discoveries, even during normal science, since we do not regard the outcome of experiments as fixed by our beliefs; things are instead the other way around. More crucially, the possibility of anomalies would be a mystery. If the world is just what the paradigm says it is, then the world should not force upon us beliefs that contradict the paradigm.

There are nonetheless other forms of idealism or related views from which Kuhn is not so far distant. Berkeley, for example, argued that there is no material world. What we think of as things are just constructions of ideas that exist in the mind. But those ideas are typically not subject to the will and so we may have ideas that conflict with our non-perceptual, theoretical beliefs. This raises the question, where do the ideas come from? For Berkeley this key role is played by God. For other philosophers the role may be played by something that is part of the world but essentially unknowable. Such is Kant's view and because there are unknowable "things-in-themselves" his view is not strictly an idealist one. Nonetheless, the attributes given by Kant to things-in-themselves are so few that others following in his wake, most notably Hegel, suggested that we could do without the things-in-themselves, thereby returning to a purer form of idealism.

Kuhn's views have significant affinities with Kant's, which he later acknowledged. But there are one or two places where his view seems almost Hegelian. Kuhn's world has two components. It has an unchanging aspect that in some sense is the ultimate explanation of our perceptual experiences, and it has a aspect that changes in response to paradigm shifts. This is the way we perceive things.

Before looking at this in more detail, we should be aware of part of Kuhn's motivation for a story of world changes. At the outset of this chapter I articulated the empiricist conception of observation and perception, which Kuhn associates with a tradition going back to Descartes.[35] According to this tradition, says Kuhn, the mind is presented with certain "data", the given. The nature of the data is independent of previous experiences and beliefs, and of the conceptual apparatus of the subject. Prior experience and the like come into play only in the interpretation of the data, when we seek to make inferences about the world. This view allows us to define (some, at least, of) our basic concepts in terms of the basic elements of experience. Since such elements, the data, are given to us uninterpreted there is no danger that changes in belief will affect these basic concepts. Furthermore, if such basic concepts are shared, then any conceptual difficulties arising because of divergence of belief may be resolved by translating the offending concepts into the shared basic concepts. Hence there may be the possibility of a theory-neutral observation language.

In this view we may regard the data – perceptual experiences or "sensations" in Kuhn's vocabulary – as constituting an interface between the mind and the world that has a particular epistemological role.[36] On the one hand, we can take these experiences to be in the mind as they are a key part of our mental life. On the other hand, their content is determined outside the mind. (It may be determined by an "external" physical world, perhaps by God. The content available may be limited, for example, by the perceptual powers available or by sensory orientation.)

While it is helpful to talk about a Cartesian tradition, it should be noted that this tradition has important strands that do not fit Kuhn's description exactly. An important part of the tradition allows for more than one kind of experience and hence more than one kind of corresponding concept. Associated primarily with Locke, but also to be found in Descartes and Galileo, is the thought that there two kinds of quality of things – primary qualities and secondary.[37] Primary qualities produce in the mind ideas (perceptual experiences or Kuhn's sensations) that resemble those qualities. The precise nature of the perceptual experience of a primary quality may be determined by a geometrical calculation; thus one length may look shorter than another by virtue of being farther away; the rules of perspective may be used to predict this. In contrast, secondary qualities produce ideas that do not resemble them. What we call the green colour of grass is just a certain arrangement of atoms in the grass that is able to produce in a perceiver the characteristic experience of green; that arrangement of atoms bears no relation to the "greenness" of the experience. The question then arises, where does that special quality of the experience come from? Locke regarded this as a product of the mind itself; or, rather, it is a product of the mind acting in concert with the physical nature of the light incident upon the retina. Locke held that different minds might produce different experiential qualities even when the eye is irradiated with the same light. My mind might be caused by the light from grass to give me an experience that I call "green", but the same light might cause your mind to have an experience I would call "pink". Of course, as long as the responses to the different wavelengths, intensities and saturations differ between us in a systematic fashion, we will find that we do not disagree about what things we are willing to call "green", "pink" and so on.

We will return to this complication of the Cartesian tradition shortly, since in fact it has some significant affinities with Kuhn's own views. For the time being we shall work with the simpler version, where the mind is entirely passive as regards the quality of experiences it has (at least for any given bodily orientation and degree of perceptual capacity). We may think of the material, or "external", "world" as that which exists independently of the mind. In this definition of "world", the world includes everything that may determine the content of the mind. If we compare this view with Kuhn's we see that according to the latter the "world" in this sense does *not* include everything that determines mental content. The fact that an individual's mental history, past experience in particular, affects aspects of perception means that the world (as-independent-of-the-mind) leaves out some content-determining factors. This means that we might choose to expand the concept of world to include the latter. To do justice to the facts of perception we need two world concepts. The first Kuhn sometimes calls "nature"; following Hoyningen-Huene we may also call this the "world-in-itself".[38] The second world concept, that which covers the things that help fix the nature of experience, is Kuhn's "world" in the sense that allows for world changes. This Hoyningen-Huene calls the "phenomenal world".[39] The mind and the phenomenal world overlap. Thus certain mental changes may result in changes in phenomenal world (but without requiring any change in the world-in-itself).

Now let us return to the amended version of the Cartesian tradition that allows for a distinction between primary and secondary qualities. In reaction to Locke, Berkeley argued that no satisfactory distinction between primary and secondary qualities could be drawn – our experiences of primary qualities no less than our ideas of secondary ones should be regarded as originating in the mind. In a similar vein, Hume held that many of our beliefs, and the concepts we use to express those beliefs, have their origins in the mind rather than in the world. So, for example, he argued that our notion of causation, which we unreflectingly think of as applying to relations among things, really stems from the mind's expectation of seeing one thing (e.g. a broken window), having experienced another (a flying stone), an expectation induced by the habituation of seeing the former follow the latter. Thus we should think of causation not as a feature of the world-in-itself, but rather as something we believe we see in the world but which is really brought about by the mind. More generally, Hume argued that there is no good argument that shows that there is a world-in-itself at all. As an argument, scepticism about the external world is irrefutable. But belief in its existence is ineradicable, even for those who find the sceptical argument intellectually compelling. Belief in an external world is thus also to be explained in terms of a disposition of the mind.

Kant may be thought of as having taken the Berkeley–Hume line of thinking to a systematic extreme. For Kant no feature of our thought can be said to originate in, or correspond to, the world-in-itself. (Kant himself talks of "things-in-themselves".) We can know of no property of any thing-in-itself, nor do any of our concepts latch on to any feature of a thing-in-itself. We might be tempted to think that if things-in-themselves exist then surely we may say that they exist in space and time. But Kant denies even this. Our capacity to have experiences Kant calls "intuition" (not to be confused with our modern, everyday notion of intuition). Space and time he regards as *forms* of intuition. That is, they characterize the way the mind organizes its experiences and so should not be thought of as external dimensions in which the things-in-themselves exist.

Paul Hoyningen-Huene draws parallels between Kuhn's views, as he reconstructs them, and Kant's. The key parallel is between Kant's thing-in-themselves and Kuhn's world-in-itself (so named by Hoyningen-Huene because of the parallel with Kuhn). Both are unknowable. Both exist beyond our possible experience.

The second parallel is that what the mind does experience is generated by the mind. But here there is an important difference between Kant and Kuhn. According to Kant the way in which the mind generates and organizes experience is a fixed part of our natures, while Kuhn's view is that the mental contribution to experience can change; a paradigm shift will lead to a different set of perceptions. What for Kant is static is for Kuhn mutable – Kuhn is a dynamic Kantian.

Possibly another difference is that Kant's things-in-themselves are completely unknowable and have no describable connections with our experiences, while Kuhn seems in *Structure* to be rather less extreme (though rather later this difference disappears). Hoyningen-Huene argues convincingly that according to Kuhn's view we should be able to say some things about his world-in-itself. We can say that it is differentiated (i.e. it is not just one homogeneous stuff but has distinct bits), and that its parts exist in space and time. The reason we can say such things is that Kuhn believes there must be such a world in order to account for the fact that adherents of the same paradigm by and large share the same experiences.[40] This shows we can say more. The world-in-itself is causally responsible for our experiences (in conjunction with the paradigms instilled in our minds), even though we cannot say how.

Perceptual experiences are what Hoyningen-Huene calls "subject-sided moments". The distinction between the world-in-itself and the phenomenal world is that the latter consists of the former plus subject-sided moments. Hoyningen-Huene is thereby able to explain why Kuhn says such things as "though the world does not change with a change of paradigm, the scientist afterward works in a different world".[41] The first use of "world" refers to the world-in-itself, while the second indicates the phenomenal world. It is clear that Kuhn does not countenance idealism, and that he does not think that the only legitimate sense of "world" is that in which it changes as a result of paradigm change.

This Kantian picture may be elaborated in more than one way, corresponding to different ways of interpreting Kant himself. One the one hand there is an interpretation that puts Kant firmly in the Cartesian or even empiricist tradition. In this view Kant holds that all we are directly aware of are our ideas. The "phenomena" are purely mental. Such a view has much in common with a sceptical version of Descartes or Locke. One of the reasons that Berkeley adopted idealism was his conviction that Locke's view leads to scepticism. Locke says that the primary qualities of things resemble the ideas they produce in our minds, but that secondary qualities do not resemble the ideas they produce. But how, says Berkeley, if all our knowledge is mediated by ideas, do we independently get to the qualities themselves to see whether or not they resemble the ideas that they produce? In some respects Kuhn is closer to a Lockean who accepts Berkeley's charge of scepticism than to Kant, since Kuhn accepts that the subject's mind is in causal contact with the rest of the world. The difference with a sceptical Lockean or Cartesian would be the same as the difference with Kant, that the mind's contribution to its experiences is not fixed but is changeable. It is this belief in changeability that Kuhn regards as constituting his break with Cartesianism and the subsequent empiricist tradition. Yet it needs to be noted that this break requires retention of significant Cartesian–empiricist doctrines, most especially the claim that there is an epistemologically deep divide between the subject's ideas (or sensations, perceptual experiences, observations etc.), whether changeable or not, on the one hand, and the rest of the world, on the other.

A more sophisticated view denies that we are aware only of our ideas. Instead we are in direct perceptual contact with the observable portion of the world. Even so, there is no knowable division of the world into different things or different kinds of thing. Yet we perceive it as if it were so divided. Furthermore, says this view, the differentiation of the world into things and kinds is an inalienable part of perceptual experience. It is not, as empiricists have thought, that the basic parts of experience are unconceptualized data that the mind then interprets as showing the existence of things and kinds.[42] Rather, the process of conceptualization takes place earlier, before the subject experiences anything. Experience comes ready conceptualized. The duck-rabbit is useful again as an illustration, for the diagram is not inherently either a duck or a rabbit. When one sees the diagram as a rabbit one sees the diagram directly, but its "rabbitiness" is supplied by the mind. The Kantian view, in this interpretation, is that all experience is like this. So there is nothing analogous to seeing the diagram neutrally, as neither duck nor rabbit. For Kant the contribution of the mind is a fixed part of human nature. For Kuhn it is not fixed: it is a variable feature, one that can change with a revolution. Employing the analogy once more, for Kant one could only ever see the duck-rabbit as a rabbit; for Kuhn a change in conceptualization, like a gestalt switch, is possible. Note that when I have been talking about the conceptualization involved in experience I mean a deep kind of conceptualization that is perceptual in nature. It is the sort of thing that gestalt psychologists are concerned with when we see significant form in the world. It is not the shallower sort of conceptualization that is achievable merely by a change in vocabulary.

To conclude this section I will comment on later developments in Kuhn's thinking. As he became more conscious of similarities with Kant their differences (over whether the world-in-itself can be said to be differentiated, causally efficacious, and existing in space and time) became less marked. In 1979 he avows himself a dynamic Kantian but is silent on the forms of intuition (space and time).[43] In 1990 he says explicitly that the "source of stability" is *outside* space and time. He also says that it is like Kant's thing-in-itself – ineffable, undescribable, undiscussible.[44] At other points in his later writing Kuhn seems to want to go *beyond* Kant. Partly motivated by concerns about the very concepts of truth and reality, Kuhn suggests that we could do without the world-in-itself altogether.[45] When he does so Kuhn looks rather more like Hegel than Kant.

Hegel amends Kant in two ways. First, he denies the existence of things-in-themselves; secondly, he says that the organization of ideas and thought is not fixed but undergoes development. Hegel's "Absolute Spirit" (the world of thought) transforms itself in a dialectical, historical process. The key feature of this process is that the "thesis", such as a particular idea in history, gives rise to a second idea, the "antithesis", in conflict with the first; the tension between the two is resolved in a new idea, the "synthesis" that in some way combines the thesis and antithesis. Removing the world-in-itself from Kuhn's phenomenal world matches the first step from Kant to Hegel. This residual phenomenal world of Kuhn's undergoes historical development like Hegel's Absolute but unlike Kant's phenomenal world. Hence we have the second step from Kant to Hegel. Furthermore, the kind of development Kuhn gives us is just that described by Hegel. A paradigm is just the sort of thing that can be a Hegelian thesis. Correspondingly, the antithesis will be a crisis-evoking anomaly. Importantly, the antithesis must be generated by the thesis itself, and this is what Kuhn tells us. Anomalies are not simply recalcitrant experiences that might have come from just anywhere. On the contrary, Kuhn asserts that they can come only from the existing paradigm. Sometimes anomalies are puzzles set by the paradigm that cannot be solved (such as explaining the precession of the perihelion of Mercury in Newtonian astronomy). In other cases they are observations or experimental results that conflict with expectations generated by the paradigm, and even in those cases those observations will have been made only in the context of carrying out normal science under that paradigm (such as Roentgen's anomalous X-rays discovered while using the cathode ray tube in routine experiments). It is conflict between paradigm and anomaly that brings about a crisis and ultimately a revolution that replaces the old paradigm by a new one. The new paradigm is a true synthesis of the old paradigm and the anomaly since an acceptable replacement paradigm must as far as possible preserve the puzzle-solving power of its predecessor while also solving the new puzzles generated by the anomaly.

World change and perception

Having seen how Kuhn's world change thesis might be understood, we should now ask whether there is any reason to believe it. First, we might ask whether there is anything in the Kantian views it resembles. On the empiricist interpretation of Kant, our ideas (perceptual experiences) are the only direct objects of awareness. This "sense-datum" view has little credibility in contemporary philosophy and there is no need to go into the reasons here. It is nonetheless worth remarking that were this Kuhn's view it would show a very strong element of empiricism in his thought. Rejection of the independence thesis means that Kuhn thinks that the nature of experience can change as a result in changes of other mental states (such as theoretical belief). But this is consistent with the thought that all that is known to us is sense experience. The (deep) conceptualist interpretation does not go in for sense data as the direct objects of sensory awareness – what we experience are independently existing things, although our experience comes conceptualized in a manner that depends on the mind. This view gets off to a much better start than the empiricist picture, but in Kant's hands there is a scepticism that is implausible, especially from a post-Darwinian perspective. We are supposed to think that we have no reason to believe that our conceptualization matches the way things in the world actually divide up. Imagine a creature whose conceptualization completely failed to mirror the world's structure. Such a creature would be utterly unable to negotiate its way around the world. Its survival would be short-lived. The fact that we and other creatures by and large get along all right suggests not only that we do have conceptualizations that match the world's divisions, but also points to the reason why we should expect this to be the case. Although Kant took conceptualization to be a product of human nature, he did not have the benefit of knowing that human nature, this fundamental part of it at least, is itself the result of millions of years of evolution. Creatures whose conceptualization matches the world well are more likely to survive and to pass their genes on to subsequent generations than creatures who conceptualize poorly. Indeed Kuhn notes the evolutionary benefits of accurately detecting differences in kinds of thing.[46] The early Kuhn, I believe, did not share Kant's general scepticism about our ability to connect with at least some perceptible kinds. He may consistently hold that for the most part our conceptualizations are accurate, carving nature at its joints, while also believing that there is room for variation. Kuhn's remark that "the scientist can have no recourse above or beyond what he sees with his eyes and instruments" suggests that he thinks that many things in our environment are perceived to some degree as they actually are.[47]

Let us now turn to the world change thesis itself. The thesis is that a change in paradigm brings with it a change in world, where "world" is understood as being the phenomenal world. In particular it is the "subject-sided moments", the perceptual experiences, that change. This part of the discussion will not depend much on whether one prefers the sense-datum interpretation or the conceptualist interpretation of Kuhn's Kantianism. It will be important, however, that we bear in mind that the conceptualization involved in that interpretation is a deep conceptualization. When I discussed paradigms in detail we saw that Kuhn likes to think of the operation of paradigms-as-exemplars in visual terms or at least in terms of visual analogies. While this is fruitful in important respects it remains true that exemplary puzzle-solutions often have a very small or non-existent visual component. Indeed this is typical in the theoretical and mathematical sciences. Acquiring a new paradigm of one of these kinds is a matter of acquiring theoretical understanding and non-visual beliefs. So the question is, can worlds change as a result of changes of a non-visual, theoretical kind? Given the conception of "world" in play, this amounts to the question, can perceptual experience change as a consequence of theoretical change?

Our study of Hanson and Kuhn on perception suggested that it does not. We saw that there might be some marginal effect on perceptual experience or an effect operating through self-deception. But this is not enough to show that perceptual change (and so world change) is a characteristic of even revolutionary theoretical change.

We saw also that Hanson and Kuhn elided genuine perceptual change with changes in the concepts used to characterize perception. This allowed them to see perceptual change as far more common than it really is. Consequently Kuhn is inclined to see world change as being easier and more common than it is. But the sort of conceptualization involved in a mere change in concepts, such as learning the term "X-ray tube", is the shallow kind. It is not the deep kind of conceptualization that is in play when we are thinking in Kantian terms. Phenomenal worlds cannot be changed simply by a change in concepts. So, even though theoretical change can bring with it a new set of concepts or a shift in concept usage, that is not itself sufficient to generate a world change. We could weaken the world concept to make it dependent not on genuine perceptual experience but merely on concepts: a world not of genuine perceptual experience but a world of Hanson's weak perceptions. A change merely in concepts leads to no more than a change in what people say and believe they see. Then the thesis that the world changes with a change in paradigm amounts to the empty claim that we change what we believe we see when we change what we believe there is.

As far as theoretical changes in paradigms are concerned, the conclusion is either that we take the subject-sided part of the world to be constituted by perceptual experience taken in a strong way, in which case the thesis is false; or we take it in a weak way in which case the thesis is little more than trivial. What about changes to the visual parts of paradigms? The problem here has already been raised. Paradigms frequently do not have a significant visual component. Kuhn gives no examples of paradigm shifts that are visual to the same extent as the Bruner and Postman experiments. Those experiments and much other research show that visual learning moulds our perceptual experiences. Learned contingent generalizations can get built into the nature of experience. Correspondingly we should expect changes in visual experience to follow from changes in the patterns of visual learning; Kuhn's case of a child learning to distinguish ducks, geese and swans exemplifies this.[48] But the sorts of pattern and generalization involved are usually low level and are not characteristic of scientific paradigms. Nonetheless it is clear that *some* such cases play a role in science, often associated with learning to use a new piece of equipment. Learning to use a microscope is genuinely learning to have new kinds of visual experience, as is learning to read bubble-chamber photographs or X-ray plates. But a case only for a limited world change thesis may be built upon such instances. Again there is no argument that gets us from consideration of what is involved in learning to use a microscope to the thesis that Lavoisier and Priestley saw different things when looking at a jar of oxygen.

World change and intuition

The preceding section shows why I do not think that there is much basis for a substantial world change thesis that sees world changes in perceptual terms. In this section I consider a different account of what Kuhn might be driving at. This does involve a liberalization of the "world" concept to allow for non-perceptual elements, but does not go so far as to permit every theoretical change to bring about a change in the world. My suggestion draws upon associationism in psychology. Human minds, even those of the incurious, are adept at making connections and associations among things. An event may be recognized as being the cause of another or as instantiating some more general phenomenon. Some such connections are made intuitively by us all. The connection we make between dropping the plate and its smashing is an obvious case. In other cases these connections are made in the first instance by inference. That John's cancer was caused by his smoking is something we know as inference from certain post-mortem evidence and scientific knowledge. Similarly, that AIDS is a viral disease, or that blue whales are mammals but whale-sharks are not, are facts first known by reasoning from the results of scientific investigations. However, even in the cases where the beliefs that support the connections are theoretical, those connections can become a standing part of our mentality. Once these connections have become internalized we may regard them as the associations studied by psychologists and students of the mind since Hume and Bain.

Learned associations may become akin to untutored intuition. The marine biologist need not infer that of the creatures she sees in front of her that one is a mammal and the other is not. Furthermore, the associations may be strong enough that when she thinks of the blue whale she inevitably thinks of it as a mammal. The shift from something's being an inferred connection to its being a quasi-intuitive association may be explained by its becoming wired into the brain in a connectionist manner, as described in Chapter 3.[49] Such associations colour our cognitive thinking about the world in much the same way as things have emotional associations for us. Hearing the name of a loved one or revisiting a scene of an unhappy incident may unavoidably trigger emotional reactions; cognitive reactions may be triggered in just the same way. (Hume thought that such triggering is almost all that is to be said about the nature of causation.)

If we define an individual's "world" as including his or her quasi-intuitive associations then a world change would come about when a paradigm shift leads to the breaking of some such associations and their replacement by new ones. This sort of world change is illustrated by the shift in our feelings about and attitudes towards higher primates as a result of the Darwinian revolution. In 1800 the view was widespread that humans are different in important respects from all other animals; humans have souls, the animals are automata; humans were created in a one special act of creation, all the other animals were created in a separate divine act; if evolution has occurred at all, it is animals that have evolved, not humans. A feature of the Darwinian revolution was that it denied these differences. Although only a tangential feature of the overall picture presented in *The Origin of Species* it was nonetheless seized upon as a locus of dispute among the pro- and anti-Darwinians (for example, in a famous exchange between Bishop Wilberforce and T. H. Huxley). Are we or are we not descended from ape-like creatures? Are the higher primates our cousins, or are they completely unlike us in origin and nature? The victory of Darwinism means that we now make an association where our pre-Darwinian ancestors made none. We think of chimpanzees and gorillas as being like us; we are inclined to sense a closer kinship between humans and these species than holds between primates and fish or dogs. Our ancestors would have had quite the opposite thoughts. These thoughts are not just theoretical beliefs that we espouse when called upon to do so. Rather, I suggest, they have the quasi-intuitive nature I have been discussing. Although we have them thanks to learning about evolution, the connections we make, such as sensing a kinship with primates, is an association we make involuntarily. It inevitably colours our thinking about and so our interaction with primates.

I have looked at several ways of understanding Kuhn's world concept: foremost was the world as consisting of perceptual experience (including deep conceptualization); we also considered worlds as being made up of weak perception (and shallow conceptualization) and of theoretical belief. The associations I am talking about are none of these but are not completely independent of them. Importantly, they can be brought into existence and changed by theoretical beliefs much more easily than perceptual experiences. This means that, as briefly exemplified by the primate case, a change in theoretical paradigm can bring about a world change, in this sense of "world". On the other hand, such associations are not *equivalent* to theoretical beliefs, since it is not necessary that possessing a theoretical belief establishes a quasi-intuitive connection. Hence it is not trivial to say that a change in belief may lead to a world change. The associations in question are not always associations to or from perceptual states, so it cannot be that all changes in these associations involve perceptual change. But some associations can be activated by seeing things, such as when seeing a chimp disposes us to think of its behaviour in human terms. Are our perceptual experiences affected by these associations in such cases? Could perception change as a result of a change in associations? I think it might, and although this effect is no part of the current version of the world change thesis, it might help in part explain Kuhn's thinking in perceptual terms. I have already suggested that it may not be unreasonable to think of our being able literally to see causal relations, and where we do, that is a result of our associations at work. I conjecture that associationism also provides the best explanation of what "seeing-as" amounts to; we see the chimp as akin to us. One thing that such associations do is to make certain descriptions natural or salient for us. And so a change in associations will mean a change in associated descriptions and so may lead to a change in intensional perception. Hence Hanson's weak seeing will be affected by changes in associations (but not every change in weak seeing is brought about by a change in associations). The suggestion is that the right way to understand the difference between Galileo and Tycho at dawn is not to think of them as having different perceptual experiences, for they do not. On the other hand there is more going on than simply that they have differing theoretical beliefs. Rather, Tycho's theoretical beliefs have become wired into his cognitive structure so that on seeing the "rising" Sun he intuitively makes the connection to the thought of the Sun as a moving satellite of the Earth, while Galileo's different associations cause him to think of the Sun's coming into view as resulting from the rotation of the Earth (a layman may have established no associations of this kind and so no corresponding thought enters his head at all). If this is right then we have a substantive proposal for what world change amounts to, which does not, implausibly, require change in perceptual experience, nor is so weak as to be simply identical with change in vocabulary or theoretical belief. It incorporates an interesting hypothesis about the nature of the mind that is of a piece with Kuhn's view that scientific thinking is not all rule-governed but involves connectionist-style explanations, a hypothesis that, I suggest, has considerable prima facie plausibility. Furthermore, as a suggestion as to what Kuhn was seeking with his world change thesis, it links with his sense that the use of perceptual words is, on the one hand, natural and not purely metaphorical yet, on the other hand, is not a central use of such words. We do talk of a change in perception meaning a change not so much in sensory experience but more in how a person feels and is apt to think about some matter. This is a natural employment of the term that is best explained by the existence of quasi-intuitive associations.

Social constructivism

We have seen that Kuhn is no idealist; he thinks that there is a mind-independent world-in-itself. But he is a sceptic; he thinks the world-in-itself is unknowable. Nonetheless our theories are often about the world-in-itself. And those theories change. Scepticism suggests that the explanation of why a particular new theory is adopted cannot be that the evidence makes that choice rationally inevitable. So what instead explains why some theories are preferred to other possibilities? One answer is that it is social forces that explain the details of theory choice. This answer belongs to a view known as "social constructivism" (and sometimes also "social constructionism").

A weak version of social constructivism fully admits the existence of a reality about which scientists theorize and have beliefs, but simply denies that it has a determining influence on the theories we choose. It is thus in part an error theory – scientists are largely mistaken in thinking that they are making choices rationally dependent on the evidence (and thence on the world itself) that lead them to the truth and give them knowledge. Choices are instead explained by the political allegiances of scientists, class interests, indoctrination, nationalistic sentiment, social relations with other scientists, desire for professional advancement, and so on.

A stronger version of social constructivism regards the positing of an impotent reality beyond the reach of theory as pointless. Instead it takes to be "real" only what is asserted by contemporary theory. This is no longer an error theory, since what scientists believe is, on this view, true. If "true" and "real" are terms taken seriously then this view becomes a social version of idealism. Reality, the world, is just what scientists believe it to be. Often, however, "truth" and "reality" are not taken seriously. The constructivist either uses the terms rhetorically or regards their use in everyday and scientific language as essentially rhetorical, signalling solely the speaker's or society's approval of the proposition in question. Now strong social constructivism has moved away from idealism and closer to weak social constructivism, except that it leaves out mention of the existence of the largely unknowable reality behind theorizing and experience.

Because of the world change thesis, Kuhn is often regarded as a social constructivist. Is he one, and if so of what kind? For precisely the same reasons that he is not an idealist he is also not the idealistic kind of strong social constructivist. Indeed, when he explicitly acknowledges the existence of the world-in-itself he is not even the rhetorical kind of strong social constructivist. That said, as we shall see, Kuhn does in the 1969 Postscript says several things that point towards the rhetorical view. There he rejects the notion of truth in general, while allowing a paradigm-dependent use of "true".

Is Kuhn a weak social constructivist then? Adherents to the weak variety can differ about the degree to which social factors influence scientific belief. At one extreme one could hold that social or social-cum-psychological factors are the exclusive determinants of belief. At the other extreme one might think that genuinely scientific factors like experimental evidence and methodological rules play the major role and that social factors are more limited in their operation. If Kuhn is a weak social constructivist, then he is at the weaker end of weak social constructivism. He mentions only scientific factors at work in the operation of normal science. It is only with regard to revolutionary science that Kuhn cites non-scientific causes of theory choice, such as nationalistic sentiment. In fact, of the influences on revolutionary theory choice that Kuhn mentions in *The Structure of Scientific Revolutions* only nationalistic sentiment looks genuinely social. He mentions idiosyncrasies, personality and psychology, but these relate more to the individual than to society. He also discusses the aesthetic appeal of theories and the reputation of the innovative thinker as grounds for theory preference. But both of these may be regarded as intra-scientific. For it is the content of the theories that makes them aesthetically attractive or not. And it is presumably the scientific reputation of the innovator that Kuhn thinks is influential.

Kuhn seems a very weak social constructivist indeed. But there is a response that seems to resist my conclusion. Is it not the case that the factors I have called "intra-scientific" are nonetheless socially inculcated? Does not Kuhn emphasize the role of professional socialization in ensuring that young scientists employ the current paradigm?

The point may be reinforced by pointing out that Kuhn, along with many other philosophers of science (constructivists included), emphasizes that the evidence plus rules of inference fail to determine a unique answer to a question of belief revision or extension, even in the context of an accepted paradigm. So when scientists do fix upon an answer, there must be some additional factor in play. This might be overtly social. In normal science, according to Kuhn, the additional factor is the intuition generated by the paradigm. The paradigm is itself an object of consensus, forced upon scientists in their training. In this response there are at least two issues that merit comment. First, the response assumes that if evidence plus rational *rules* are insufficient to determine belief, then whatever additional factor is required will be grist to the constructivist's mill. This in turn assumes that scientific rationality must be rule-based. That assumption is central to logical empiricism. The operation of paradigms is not rule-based. Does that show that paradigms are irrational? Or does it show that rules are not essential to rationality? I suspect that because both critics and supposed supporters of Kuhn have accepted this empiricist assumption, they have all seen his emphasis on paradigms as undermining rationality. But it seems equally plausible as an interpretation of Kuhn (as I was arguing in Chapter 3), and also as a philosophically sounder position (as I will argue in Chapter 6), to hold that rational progress in science does not require belief change to be fixed only by rules. In which case choice of puzzle-solution in normal science may not be rule-based, but it may nonetheless be rational and not explained by non-scientific forces. The second, related point is that the response under discussion also assumes that belief-forming processes that are social in nature or are socially inculcated cannot be rational. This is a contentious claim – and I shall contend it too, also in Chapter 6. For the time being I shall remark that individualism about rationality is also an empiricist assumption, one that reflects an internalism about epistemology, the view that if a subject has a belief that is rational (or is knowledge), then the subject must be in a position to know that the belief is rational (or is knowledge). An improved, externalist view of epistemology rejects this and so makes room for social rationality and social knowing. To sum up, if rational belief can be acquired by non-rule-governed, social processes, then Kuhn's account of paradigms does not exclude the possibility of rational belief. And so to the extent that social constructivism stands in contrast to explanations of belief formation that show them to be rational, Kuhn need not be seen as a social constructivist.[50]

Stimulus and sensation

In the Postscript 1969, Kuhn expands on the relationship between the world-in-itself and the phenomenal world, employing the terminology of "stimulus" and "sensation". Sensations are our experiences – they are included in that part of the phenomenal world that goes beyond the world-in-itself, they are "subject-sided". Stimuli, on the other hand, are part of the world-in-itself, they are object-sided: they are the "worldly" causes of sensations. The same sensations may be brought about by different stimuli (as in certain illusions), while, more importantly for Kuhn, the same stimulus may result in different sensations. This is because the stimuli are not sufficient to account completely for the character of the sensations; crucially the mind, programmed by exposure to exemplars, has an influence as well. Thus persons belonging to different paradigms may receive identical stimuli yet have different sensations as a result.

Stimuli are those bits of the world that impinge on the surfaces of our sense organs. "If [two people] could put their eyes at the same place, the stimuli would be identical."[51] Thus we might infer that stimuli include the light rays that strike our retinas, the sound waves incident upon our ears, the chemicals that excite our taste buds, and so on (but Kuhn does not actually say as much). On the other hand, he does explain that there is much neural processing that goes on between the stimulus and the resulting sensation. This is where exemplars in our scientific education get to play their role.[52] Talk of sensation suggests that of the interpretations of Kuhn's world concept we have discussed, the world-as-constituted-by perceptual-experiences is the right one. The problem with the world change thesis is then that although Kuhn has shown that differences in perceptual history may cause people with the same stimuli to have different sensations, he has not shown that differences in theoretical belief can have that effect.

The fact that the same stimuli may give rise to different sensations suggests to Kuhn that stimuli cannot themselves be perceived.[53] The view of stimuli as including incident light and sound rays supports this. Our knowledge of stimuli is, says Kuhn, "highly theoretical and abstract".[54] It is comments like these and the remarks on neural processing that give rise to a considerable tension in Kuhn's position. At the simplest level Kuhn first tells us that we cannot know much about the world-in-itself and then goes on to tell us much about it – that we know that it includes stimuli which get processed neurally. This is knowledge that is "highly theoretical and abstract". Hoyningen-Huene characterizes this as a tension between two standpoints.[55] The one standpoint Hoyningen-Huene calls the *critical epistemological* standpoint. This standpoint is neutral between different phenomenal worlds, i.e. it does not buy into any one paradigm or set of theories, but recognizing how these may change rejects knowledge of any of them. The other standpoint is the *natural* standpoint. This standpoint is happy to adopt whatever best theories current science has to offer. The critical standpoint will thus minimize our knowledge of stimuli as object-sided entities, while the natural standpoint will bring science in to tell us their nature, how they interact with our neural structures and so on. Hoyningen-Huene is right that Kuhn's position seems to be in an unhappy tension between a willingness to have science tell us about stimuli and the claim that object-sided things like stimuli are unknowable.

This tension is conceivably resolvable. Kuhn could argue that the critical standpoint is critical and sceptical of scientific theories only when those are understood in a realist sort of fashion, as being true or close to the truth. If our acceptance of science is not conceived of in this way, but instead as our best shot at solving certain puzzles, then the critical standpoint ought not preclude our operating with the current paradigm. After all, Kuhn is himself a theoretician, he seeks to solve a puzzle about the nature of scientific change and presents a hypothesis concerning it. There is no reason why he should not make use of current science is helping him solve his puzzle.[56]

The fact that Kuhn does not give this answer suggests that from some point in the 1970s he does not regard his work as having the same status as the science it discusses – he is not solving problems in the theoretical history of science but is instead telling us something that is potentially worthy of belief. The belief is a philosophical not a historical or scientific one. The critical standpoint requires little or no knowledge that is seriously theoretical. For that we need only the existence of something in the world whose causal impact on our sense organs is roughly the same for everybody, at least for those similarly positioned. And this says Kuhn we must accept "on pain of solipsism".[57] A claim we accept because we deny solipsism is not a theoretical conclusion given to us by a particular science; if anything it is a pre-scientific philosophical proposition.

As I mentioned earlier, Kuhn's thought later becomes more explicitly Kantian and sceptical about the world-in-itself. Correspondingly, appeals to the findings of research in psychology, neuroscience and artificial intelligence are dropped. Kuhn rejected the empirical approach he earlier took and sought to re-establish his conclusions from "first principles". In effect, given the choice between the critical and the natural standpoints, Kuhn opts firmly for the former. This seems to me to be an important turning point in Kuhn's thinking – and in my view he took the wrong road. As I shall explain in Chapter

6, the most important and radical departure in epistemology has been the development of externalist, naturalized epistemology. This approach, which requires a deep rejection of Cartesian and empiricist epistemology, would endorse the natural standpoint. By adopting the critical standpoint Kuhn has shown that however much he sees himself as reacting against Cartesianism he is nonetheless fundamentally committed to its basic outlook.

Kuhn and constructive empiricism

To conclude this chapter I shall introduce a philosophy of science that has surprising affinities with Kuhn's approach. Bas van Fraassen's constructive empiricism says that our theories are radically underdetermined by the data.[58] There are many, typically infinitely many, possible hypotheses that are consistent with our evidence. Certain rational principles will allow us to reduce the range of viable hypotheses. But since any rational argument for or against a hypothesis is based on observed data, our principles of theory preference can only be sensitive to the truth of the observational portions of our hypotheses. Therefore such principles can at best only select those theories that are likely to be empirically adequate – those theories that have true observational consequences (but may have false unobservable consequences). Van Fraassen rejects as rationally ungrounded (indeed as rationally objectionable) modes of reasoning such as Inference to the Best Explanation that would allow us to pare down the viable hypotheses to a small number. We are not entitled to regard as true even our favourite, best confirmed theories if they have consequences regarding unobservables.

Hence for van Fraassen, as for Kuhn and Kant, there is an unknowable world-in-itself. Kuhn puts it in words that van Fraassen might have used, "The scientist can have no recourse above and beyond what he sees with his eyes and instruments".[59] All three are sceptical as regards hypotheses that go beyond our experience, since the evidence presented to us by experience fails to determine the truth of such hypotheses. If we were to think that the evidence plus rational rules and principles of inference, such as logic and mathematics, could determine just one or a handful of viable hypotheses, then we would have a clear scheme of explanation for the history of theory choice. Say we want to explain why theory h_i was adopted. A simple explanation might go like this. Investigation, observation and experiment produce a body of data. The data rationally constrained the available hypotheses to a handful – $h_1, h_2, \ldots, h_n$. Given the small range of hypotheses under consideration scientists were able to develop further tests and experiments whose outcome would decide between them. After carrying out the tests, all the hypotheses other than h_i were rejected.

Now consider the problem of explaining theory choice if van Fraassen is right.[60] Then the data plus rationally acceptable inference rules will not constrain the range of viable hypotheses to just a handful. The range of hypotheses that are rationally acceptable will be potentially infinite. Yet it remains a fact that scientists do work only with a limited range of hypotheses and indeed generally regard it as difficult to think up acceptable hypotheses rather than easy. So something in practice constrains the actual choices made by scientists. Since evidence and rational rules do not limit theory choice to the small number actually considered, some other factors must be in play. We have seen earlier in this chapter that this is where the social constructivist finds room for social and political factors which, from the Old Rationalist standpoint at least, may seem irrational. I suggested that Kuhn's conception of the paradigm also fills this gap between rational rules and theory choice. But this is not so much because paradigms are irrational but because they are not rule-governed. Indeed, van Fraassen himself notes that Kuhn's paradigms fit the role well of what extra explains the choices scientists actually do make.[61] The role of the paradigm is to constrain choice of puzzle-solution. According to van Fraassen, simplicity or explanatory power, for example, are not rational bases for theory choice. But they are the sort of consideration that will allow reducing hypotheses from many to a few. And they are also the sort of consideration generated by a paradigm.

Van Fraassen does think that there are some objectively rational principles and rules of theory choice, Bayes' theorem for example. But that does not conflict with Kuhn's views. As we have seen, Kuhn does not deny the existence or use of mathematics, logic, and other rules; he denies that they alone explain theory choice. For there to be room for Kuhnian paradigms as a source of paradigm-dependent factors in theory choice, it need only be that such objective principles are insufficient to determine theory choice. As we have seen, Kuhn's point about scientific change, whether puzzle-solving in normal science or revolutionary choice of a new basic theory, is not that it is arbitrary, nor that it employs no rational principles. Rather, it is that such principles and rules are themselves not enough to fix the choice.

The theory we have now will not necessarily always accord with future evidence. Moreover, the fact that many theories will agree with one another as regards current data but differ over observational facts not yet recorded, means that it is quite likely that current theories are not only wrong but empirically inadequate also. If the current theory is empirically inadequate this will begin to show in the production of anomalies, resulting in crisis as Kuhn describes. Now we will need to replace the core theory of a paradigm. As we have seen, there will always be a large number of hypotheses that will be consistent with the observable data. But now there will be no agreed paradigm that will provide an appropriate sense of similarity to allow selection "acceptable" hypotheses and puzzle-solutions. Therefore many more hypotheses will be available for consideration, with neither sufficient rational principles nor agreed standards to guide selection. Thus room will be made for "revolutionary" modes of promoting a theory. We might think of such modes as including social and political pressures, propaganda, the exercise of power and so on. As has been remarked, Kuhn scarcely mentions such things.[62] The factors in play are mostly similar to those operating in normal science, such as the ability to solve unsolved puzzles. It is just that these no longer operate in such a way as to fix a consensus. Less tangible, more flexible factors, such as aesthetic appeal, play a more prominent role. But these too are unable to determine a choice of theory on which all scientists will agree. During revolutionary science there is more room for disagreement. The factors (such as puzzle-solving power and aesthetic appeal) that do eventually lead to a choice of theory may take longer to reach their verdict. There is more room for extra-scientific contingencies to play a part.

In short, van Fraassen's thesis of underdetermination (by evidence plus rational rules) allows a role for Kuhn's non-rule governed paradigms in explaining puzzle-solution and theory choice. The points of contact between Kuhn and van Fraassen do not end there. A further similarity concerns the aim of science. Since scientific inference cannot be known to take us to the truth, van Fraassen denies that the goal of science can rationally be the truth of theories. Instead the aim is empirical adequacy. Scientists should want their theories to have true observational consequences. Kuhn too rejects theoretical truth as the aim of science; it is instead increasing puzzle-solving power. As we saw in Chapter 2, puzzle-solving seeks to bring the paradigm theory and reality into closer agreement. Since he is sceptical as regards the world-in-itself, Kuhn cannot mean the latter by "reality". Rather he must be referring to the subject-sided moments of the phenomenal world. Put more simply, Kuhn thinks that puzzle-solving requires getting a closer fit between theory and the results of experiment and observation. Hence for Kuhn as well as van Fraassen the goal of science is the increasing empirical accuracy of its theories.

The similarities mentioned notwithstanding, there seems to be one crucial dissimilarity between van Fraassen and Kuhn. Van Fraassen makes essential use of an observation-theory (O-T) distinction, while Kuhn is concerned to undermine the distinction. How can their views be similar or even related if this is the case? While there is at least one important difference, I shall argue that their views on the O-T distinction can be reconciled in the respects relevant to this discussion.

There are two ways, at least, in which an observation can be dependent on a theory. First, the observation may be *semantically* dependent on a theory. Secondly, it may be *epistemically* dependent.[63] The first is a claim about the nature of observational and theoretical terms that says that terms employed in the observation report have meanings that relate somehow to theories. The second is a claim about our knowledge of observable states-of-affairs that says we may have that knowledge only if some theory is true. As we have seen, Kuhn and Hanson are not exactly clear on theory-dependence but that will not affect the following discussion.

The semantic (or meaning) dependence of observation on theory is perfectly consistent with the possibility of dividing entities into observable ones and unobservable ones. Van Fraassen regards planets and stars as observable, even those not observable with the naked eye from Earth, since a suitably placed spaceman would be able to observe them. Thus supernovae are observable, even if the meaning of "supernova" is semantically dependent on theories of stellar evolution. Van Fraassen fully accepts the semantic dependence of observation on theory: "Can we divide our language into a theoretical and non-theoretical part? . . . On [this question] I am in total agreement [with Grover Maxwell, who says *no*.] All our language is thoroughly theory-infected".[64] Van Fraassen seems also to think that the semantic (or meaning) dependence of observation on theory is also consistent with dividing *truths* into those that may be known by observation and those that cannot be. It will mean that the truths of the former class may not be expressible in terms independently of theoretical expressions. But, in this view, the presence of a theoretical expression in a sentence does not make the proposition expressed unknowable by observation.[65] "This blood sample is infected with a bacterium that is a spirochete" is a sentence containing theoretical terms whose truth may be ascertainable by observation.

What then of epistemic dependence of theory on observation? Kuhn's aim is to reject the empiricist assumption of independence, that the fundamental nature of observation is unchanging, independent of either experience or theory. Kuhn does not conclude that there is no such thing as observation, nor even that there is no distinction between the observable and unobservable. Indeed his comment that the scientist has no recourse beyond what he sees with his eyes or instruments implies some such distinction. His view is only that what a subject may observe is changeable; it is paradigm-dependent. Van Fraassen is in limited agreement. It is not necessary for his point of view that what is observable be the same for all communities. He says: "what counts as an observable phenomenon is a function of what the epistemic community is (that *observable* is *observable-to-us*)".[66] What is sufficient for van Fraassen's argument is that there is a range of facts that is observational and that does not include everything which is in the domain of scientific theorizing. However, he does differ with Kuhn on the question of whether what is observable may change with change of *theoretical* commitments. For van Fraassen what is observable depends on the physiological features of members of the community. Since he is a sceptic about theories, he cannot allow observation to be epistemically theory-dependent, for that would infect observation with the same scepticism. But either way both van Fraassen and Kuhn accept a distinction between what is observable and what is not, and both agree that the distinction is a vague one – they disagree about what that distinction depends on.

Conclusion

The benefit of exploring the similarities between Kuhn and this most recent version of empiricism is to emphasize a theme that has been highlighted earlier in this chapter. Kuhn saw himself as making a break with logical empiricism and more generally with a Cartesian tradition of which it is a part. In truth he concentrates on undermining one aspect of empiricism, the independence thesis, while retaining many other empiricist characteristics. His scepticism about the unobserved and his commitment to internalism concerning the content of experience are hallmarks of empiricist thinking. In the next two chapters we shall see yet further signs of an empiricist legacy.

In seeking to undermine the independence thesis, Kuhn, following Hanson, rather overplays his hand. For example, the fact that many perceptual verbs are intensional leads them to think that the very nature of experience depends on the concepts the subject has available. Consequently it becomes easy to claim that theory change leads to change in perceptual experience, since theory change brings conceptual change too. In fact, I have argued, theoretical learning has only a weak effect on perceptual experience – certainly weaker than the effect of perceptual learning.

Kuhn's aim in attacking the independence thesis was to assist in undermining the logical empiricist account of theory change. How does this attack fare in the light of my discussion? Although, as just remarked, some of Kuhn's criticism missed its mark, it remains true that he successfully showed the independence assumption to be false, to the extent that perceptual experience depends quite strongly on previous visual training, if not on theoretical belief. Even so, I think that a more powerful attack on the empiricist account may be launched by questioning those assumptions Kuhn left largely untouched. As mentioned, internalism, assumption (iii), about perceptual experience is highly doubtful. And in the light of that doubt one should question whether an observation is a report of an experience, assumption (ii). Pre-philosophically we think of observations as being reports of things outside us; dropping internalism means that we cannot say that reports of experiences are infallible, and so we cannot say they contrast with fallible reports of external things.[67] Lastly, and most radically, we should question the observational basis. If our evidence is what we base scientific inference on, must our evidence always be observational? Cannot theoretical knowledge constitute evidence? Certainly we say that isomerism is evidence for the existence of atoms even though knowledge of isomerism involves considerable amount of theory.[68]

Since Kuhn thinks that theory change leads fairly straight-forwardly to perceptual change it is tempting to frame a world change thesis in terms of a perceptual conception of "world". In such a conception, world changes follow perceptual changes, which, supposedly, follow paradigm changes. But for the reasons summarized in the paragraph before last, paradigm changes do not typically lead to perceptual change, although changes to a paradigm high on perceptual content and low on theoretical content may perhaps lead to perceptual change. Nevertheless I do think that there is a more fruitful way of looking at world change that is not so closely tied to perceptual change. On this proposal a subject's "world" consists of (at least) the quasi-intuitive associations that the subject makes between phenomena. These associations can be forged by theoretical learning but are not just equivalent to theoretical belief. And they can change when theories change. This way the world change thesis comes out true, but not trivially so. Furthermore the proposal corresponds with one of Kuhn's more original but least noticed insights, that inference need not always be a matter of following a rule. The corresponding thought here is that belief need not always be a matter of conscious judgment. In both cases the possibilities averted to can be explained on the connectionist model of the mind.

Chapter 5

Incommensurability and meaning

The meaning of incommensurability

In the last chapter we saw that in Kuhn's opinion scientific revolutions bring with them shifts in perceptual experience and changes in the world in which the scientist operates. These may be regarded as aspects of the more general phenomenon of incommensurability. The general idea of incommensurability is that the existence of changes in perception, world, standards of evaluation or in the meanings of key theoretical terms undermines traditional, Old Rationalist conceptions of progress as the accumulation of knowledge or as increasing verisimilitude. We have already considered the claim that changes in perception mean that there is no stock of theory-neutral observations that may be used for theory evaluation. In this chapter we will look at the parallel claim that shifts in meaning similarly preclude the possibility of a common measure of theories and prevent us from seeing progress as a matter of theories improving on their predecessors by getting closer to the truth.[1]

The thesis that theories originating from different paradigms are incommensurable is one of the most keenly debated aspects of Kuhn's account of scientific change. The fact that this thesis has been taken to imply that such theories cannot be compared accounts for much of the controversy surrounding it. Yet it is important to be clear from the outset that incommensurability does not *mean* non-comparability; nor does it entail non-comparability in any trivial or straightforward way.[2] The notion of incommensurability is borrowed from mathematics. Consider a right-angled triangle that has shortest sides of length 1.0m and 0.75m. We know from Pythagoras' theorem that the hypotenuse has a length of 1.25m. In this case the lengths of the two shortest sides, when measured in metres are both rational numbers, and so is the length of the hypotenuse. This means that we could find some unit of measurement such that the lengths of all three sides are whole numbers of such units. Most simply we could use centimetres, giving us sides of 100cm, 75cm and 125cm. Or we could invent a new unit (the "quartimetre") such that the sides are 4qm, 3qm and 5qm. The fact that there exist such units so that all the lengths can measured in whole numbers of those units is expressed by saying that the lengths are *commensurable* – literally "co-measurable". It came as something of a shock to the ancient Greeks to discover that not all lengths found in simple geometry are commensurable. Take for example a right-angled triangle whose sides are both of length n units. Pythagoras' theorem tells us that the length of the hypotenuse is $\sqrt{2}.n$ units. Euclid records an elegant proof that $\sqrt{2}$ is not a rational number – there are no whole numbers p and q such that $\sqrt{2}=p/q$. So not only is the hypotenuse not a whole number of the units we originally used, but whatever units we care to choose, however small, if the sides are a whole number of such units, then the hypotenuse is not a whole number of those units.[3]

Thus the sides of a square are incommensurable with its diagonal. But clearly the lengths can be compared – it is trivial that the diagonal is longer than the sides. Therefore, if we take the source of the concept as providing a serious analogy, we should expect the claim about the incommensurability of theories to require the existence of a measure for theories operating within the same paradigm and the non-existence of such a measure for measuring theories from different paradigms. We should not, prima facie at least, expect the lack of a measure in the latter case to mean that the theories can in no way be compared. As Kuhn reminds us, the "lack of a common measure does not make comparison impossible. On the contrary, incommensurable magnitudes can be compared to any required degree of approximation".[4] In this chapter and the next we shall look in detail at how Kuhn applies the concept to scientific theories.

Semantic incommensurability

There are two distinct (but related) sources of incommensurability. One arises from the nature of paradigms as benchmarks for the evaluation of theories. The other stems from the role played by paradigms in establishing meaning. It is the latter, incommensurability due to shifts in meaning, that has exercised philosophers the most, yet its articulation in *The Structure of Scientific Revolutions* is sketchy. However, in his later writing Kuhn concentrated increasingly on the conceptual aspects of incommensurability – what I shall call "semantic" incommensurability – and it is this kind of incommensurability that is the focus of this chapter. We shall return to the less semantic, more epistemic kind in the next chapter. There are two aspects to semantic incommensurability corresponding to the two kinds of scientific concept Kuhn allows: empirical (or observational) concepts and theoretical concepts. Empirical concepts are concepts whose application is typically possible on the basis of perceptual experience. Theoretical concepts are those employed on the basis of the application of a theory to the world. The former are learned through the perceptible features of paradigms – ostensive definition. The latter are understood via an understanding of the theoretical aspects of a paradigm. As we have seen, Kuhn thinks both are subject to variability with paradigm change.

It is not immediately obvious why conceptual variability should be regarded as a form of incommensurability. Incommensurability is defined as the lack of a common measure. One of the roles of an exemplar is to provide a measure of the goodness of a puzzle-solution, so we can see how the lack of a common exemplar may lead to incommensurability. But concepts are not in any obvious way measures of anything. Rather, they seem to come before measurement. We choose our concepts (such as a unit of length) and then we make a measurement using them. Furthermore, we think of the results of measurement being in an important way independent of the concepts used to express the results. In the example we began with, the hypotenuse of the triangle was measured as being 1.25m or 125cm or 5qm. These do not represent different outcomes of measuring. All measurements with one of these units can easily be converted into measurements with the other units. So what is the connection between concepts and incommensurability?

The connection arises from what Kuhn takes to be the logical empiricist view of hypothesis evaluation. On this view a hypothesis is taken to have observational consequences, and the hypothesis is evaluated by seeing whether these consequences are true. This is explicit, for example, in Hempel's hypothetico-deductive model of confirmation – a hypothesis receives confirmation just in case it has observational consequences known to be true. To be precise, Hempel

thinks of the relation between theory and its observational consequences as one of deduction – we are dealing with *statements* of
observations, not the perceptual experiences themselves – and in
general the logical empiricists took observation statements as the
basis of scientific reasoning. This is where the linguistic aspect of
incommensurability begins to become relevant. Let us call a true
observation statement *relevant* to an hypothesis if it is a report that
either agrees with or disagrees with an observational consequence of
the hypothesis. From the hypothesis we may deduce the statement
"the pointer will be at 7.0 on the dial"; an actual observation that
reports "the pointer is at 8.3 on the dial" is relevant to the hypothesis.[5]

Let us now measure the quality of the hypothesis. This will be
done by collecting relevant observation statements from the results
of experiments. A large number of such statements that agree with
the corresponding consequences of the hypothesis will give the
hypothesis a high degree of confirmation. Relevant observation statements that logically conflict with the consequences of the hypothesis
(as in the case just mentioned) show it to be false. Nonetheless, if
there is a strong preponderance of confirming relevant observation
statements over falsifying ones we may still be able to accord the
hypothesis a high degree of verisimilitude (nearness to the truth).

Now let us think of co-measuring two competing hypotheses.
According to logical empiricism, two theories are competitors when
they have logically incompatible observational consequences. So
from the second hypothesis we deduce "the pointer will be at 9.5 on
the dial" which conflicts with the corresponding deduction of the first
hypothesis. Where there is such conflict, observation statements relevant to the one consequence will be relevant to the other. (In this
case the relevant observation statement falsifies the consequences of
both hypotheses.) The set of observation statements relevant to both
hypotheses will provide a (partial) measure of both hypotheses.
Furthermore, if this set is large, as we might expect for genuinely
rival hypotheses, it will provide a full measure of both hypotheses
and hence will allow for their comparison. Such a set makes possible
what Kuhn calls a "point-by-point comparison" of the two hypotheses.[6] For every observational consequence of a hypothesis we can see
whether it is true and whether the corresponding consequence of its
competitor is true.

For this co-measuring and point-by-point comparison to take place
it must be that the competing hypotheses are expressed in languages

with the same observational vocabulary. If not, observation statements relevant to the one hypothesis will not be relevant to the other. Now let us consider what happens if we consider hypotheses from either side of a shift in paradigms. As we saw in the last chapter Kuhn thinks that our perceptions change when our paradigms change. On the empiricist view, it is our perceptual experiences (sense impressions or sense data) that supply us with the meanings of our observational vocabulary. And so a shift in perceptual experience will lead to a change in the meanings of the words used to express observations and observational consequences of hypotheses. Thus the observation words used for the two hypotheses from the new and the old paradigms will mean different things, and hence observation statements relevant to the one hypothesis will not be relevant to the other and vice versa. And so we do not have a common measure of the two hypotheses nor is a point-by-point comparison available. Furthermore, it is unlikely that the paradigm shift will lead to the invention of a completely new vocabulary. Instead, old words will be used with new meanings. This means that a spurious impression may be given that there is continuity in meaning when there is not.

Although I have said that incommensurability does not mean nor lead straightforwardly to incomparability, this is an instance where a failure of one kind of comparability may ensue from semantic incommensurability. We can see that this problem is not the same as that of having different but mutually interchangeable units of measurement (the metres, centimetres and quartimetres discussed above). Kuhn thinks that the observational concepts of different paradigms are not mutually translatable. So we cannot resolve the problem of comparability by translating the observational consequences of one hypothesis into statements employing the concepts of the other:

> The point-by-point comparison of two successive theories demands a language into which at least the empirical consequences of both can be translated without loss or change. That such a language lies ready to hand has been widely assumed since at least the seventeenth century when philosophers took the neutrality of pure sensation-reports for granted and sought a 'universal character' which would display all languages for expressing them as one. Ideally the primitive vocabulary of such a language would consist of pure sense-datum terms plus syntactic connectives. Philosophers have now abandoned hope of

achieving any such ideal, but many of them continue to assume that theories can be compared by recourse to a basic vocabulary consisting entirely of words which are attached to nature in ways that are unproblematic and, to the extent necessary, independent of theory.[7]

Kuhn regards Popper as exemplifying such a view. First, Popper held the evaluation of hypotheses to be a matter of comparing their observational consequences to "basic statements", which state what scientists have observed. Kuhn takes these statements to be couched in a vocabulary of the sort he discusses in the quoted paragraph. In fact, basic statements are not attached to nature in a way that is entirely unproblematic, since Popper thinks that their truth depends on certain theoretical assumptions – a matter especially difficult for Popper since he thinks we can never know those assumptions to be true. But the matter of our knowledge of basic statements is a separate one from the issue of what they mean and the concepts in which theory is couched, and Popper does not raise any special queries on that front. Secondly, Popper's account of theory comparison does imply something related to the idea of a point-by-point comparison. Popper is a sceptic and thinks we can never have reason to believe a theory is true. But he nonetheless thinks that they can be compared for goodness and that one theory may be closer to the truth than the other – it will have greater verisimilitude. One theory has at least as much verisimilitude as another when it encompasses all the true consequences of the other, but has no false consequences that are not also consequences of the other.[8] If consequences are taken to be statements, as Popper implies, then this requires that there be some language able to state the consequences of both, making it possible that for every consequence S of one theory there is a consequence of the other that is either identical to S or to the negation of S. It is true that verisimilitude is not especially a matter of observational consequences of a theory, but they are included. And in any case Popper's account of corroboration, which is how we have access to verisimilitude, is a matter of confronting the observational consequences with basic statements generated by experimental and observational tests. If Kuhn is right about the variability of observational concepts, then this kind comparison of hypotheses from different paradigms may be ruled out.

The above quotation makes it clear that Kuhn thinks that incommensurability *does* lead to some kind of non-comparability, and no doubt this is why his critics have identified incommensurability with non-comparability. As we shall see in the next chapter, Kuhn himself thinks that theories from successive paradigms may be compared, but not for truth or truth-likeness. So we ought to understand Kuhn's attack on comparability via incommensurability as an attack on specifically truth-related comparability. Even so, we should be aware of the limitations of Kuhn's argument as it stands. First, it depends on a strong thesis of the theory- (or paradigm-) dependence of observation. As I argued in the last chapter, there is little support for a strong thesis of this kind. Secondly, there is nothing in Kuhn's account that says that hypotheses from different paradigms are *inevitably* incommensurable. As has been discussed, paradigm change need not be wholesale. Some parts may remain after a change, and so some concepts may remain constant across such a change. Thus there may be a sufficient supply of common concepts that will allow for the commensurable expression of the two hypotheses and of the observations relevant to them. Thirdly, the argument from incommensurability to non-comparability is an argument only against this kind of comparability (involving something like point-by-point comparison). But it isn't clear that this is the only kind of truth-related comparability that exists, even if it is true that logical empiricists supposed so. Returning to the mathematical case of incommensurability, it is as if we took the only way of comparing the lengths of two lines to be a matter of measuring them precisely and comparing the measurements. In that case there are clearly other ways of proving one line to be longer than another. In the scientific case one might think that Inference to the Best Explanation is a way of comparing hypotheses that does not require their point-by-point comparison but instead compares their overall explanatory goodness. Inference to the Best Explanation is thus not so obviously dependent on a theory-neutral observation language.

So far the discussion has centered on observational vocabulary, for which the incommensurability thesis is a consequence of the thesis of the theory-dependence of observation. As we shall see, Kuhn applied the thesis to theoretical language as well. Thus an attack on the Old Rationalist picture of theory comparison is extended to a challenge to the very idea of a theory being absolutely better than another by virtue of being closer to the truth. The relative nearness to the truth of two hypotheses concerns the comparison not only of their observational consequences but also of their theoretical consequences. Problems similar to those we have already examined arise for theoretical comparison if the meanings of theoretical terms shift when para-digms change. In particular, faith in scientific progress as increasing nearness to the truth may be misplaced. We might, for example, think that some advances show old theories are replaced by ones that extend their range of application or increase their precision. For example the ideal gas law $PV = nRT$ becomes increasingly inaccurate as pressure increases. The theoretical reason for this is that the equation fails to take into account such facts as the non-zero volume of the molecules themselves and the existence of forces between molecules. Hence Johannes van der Waal modified the equation thus: $(P + n^2\alpha/V^2)(V - \beta) = nRT$. Even this equation is not perfectly accu-rate in all circumstances, for example for very high or low tempera-tures. However, it is natural to see it as closer to the truth than the original ideal gas equation, which may be seen as a limiting case of, or an approximation to, van der Waal's equation, obtained by ignor-ing the usually small factors $n^2\alpha/V^2$ and β. This sort of comparison is licensed when the "P", "V", etc. in the ideal gas equation and the same terms in van der Waal's equation have the same meanings. Kuhn's contention is that it is not always the case that the theoretical revision of this sort does maintain sameness in meaning. When the revision is a revolutionary one, such as in the move from Newtonian to Einsteinian equations of motion, there may appear to be this simple relationship between the new and the old equations, but in fact this is an illusion, since the theoretical expression employed have shifted their meanings.

Kuhn, Quine and indeterminacy of translation

In the early 1970s questions of translatability, or, more exactly, untranslatability, are to the fore in Kuhn's thinking about incommen-surability, especially with regard to the question of non-comparability. Kuhn is keen to draw a distinction between untranslatability and incomprehensibility. Two languages may be mutually untranslatable or difficult to translate, even though it is possible to be an adept speaker of both and thus understand both. Understanding a language is *not* a matter of being able to translate it into one's own language.[9] (If it were, what would count as understanding one's own language?) Indeed, what led Kuhn to take an interest in incommensurability in the first place was the discovery that with an effort he was able to understand Aristotelian physics, which beforehand, from a Newtonian standpoint, seemed simply ridiculous. The appearance of absurdity was generated by the impossibility of properly translating Aristotelian ideas into a language inherited from Newton. Kuhn does remark that translation may usually be carried out *up to a point*. Accuracy of detailed import and preservation of nuance are required for perfect translation, but a full achievement of these aims would require a translation so complex as to prevent proper communication.

Kuhn linked his views on translation to Quine's indeterminacy of translation thesis. Quine considers someone investigating the language of a hitherto unknown people. The task is one of radical translation – translation without the benefit of even partial linguistic communication. The would-be translator notes that the natives tend to say "Gavagai" when they see a rabbit. While it is natural to think that the word "Gavagai" means what we mean by "rabbit", the translator could take it that they mean "undetached rabbit-part", "rabbit occurrence", or something else. Different translators could work with each of these three assumptions and draw up translation manuals accordingly. Each, says Quine, could be an accurate predictor of what the natives will say, yet would be inconsistent with the other translation manuals. Translation is indeterminate

Kuhn adds to Quine's story, supposing that in this region rabbits may change their appearance with the time of year – longer hair, change of colour, different gait and so on during the rainy season. Imagine that during this period, the natives call out "Bavagai!" Should we translate this as "wet rabbit", "brown rabbit", "shaggy rabbit" or "limping rabbit"? Or should we conclude that the native thinks that Gavagai and Bavagai are distinct kinds of animal?

Kuhn's intended conclusion is that translation is difficult because "languages cut up the world in different ways, and we have no access to a neutral sub-linguistic means of reporting".[10] We shall return to this, but before doing so some remarks on his use of Quine are apposite. Quine's indeterminacy of translation thesis has generated a lot of discussion and even then there is disagreement about what is salient to his argument. Be that as it may, it is clear that Quine's intention is to show that there is no such thing as meaning.[11] That of course would be a conclusion difficult for Kuhn to accept given that semantic incommensurability is essentially a thesis requiring the existence of a diversity of meanings. Furthermore, Quine's argument seems to show that more than one translation is available (and that the different translations make no difference). Kuhn wants to say that no fully adequate translation exists (and that the inadequacy of those that do exist does make a difference).

There are other differences between Kuhn and Quine. Quine thinks that no *possible* observations of the native speakers' utterances and circumstances could decide between the different translation manuals. One reason for this is that he has chosen what one might call "metaphysically" different translations (although Quine himself would not use this terminology). The significance of this is that whenever there is a rabbit there is an undetached rabbit part and vice versa. And so situations in which it is right to talk about the one are those in which some corresponding thing may be said about the other. Kuhn's alternative translations are not metaphysically distinct but are scientifically different. Thus, as he himself says, although we may not have enough evidence now to determine whether "shaggy rabbit" or "brown rabbit" is correct, further evidence of the right sort could answer that question (the natives' response to an albino shaggy rabbit or to a shorn brown rabbit, for example).

Kuhn says that the alternative view about what "Bavagai" means are different hypotheses. None need be right, and if the one we have chosen is wrong, difficulties may appear later in the process of communication with the natives, and we may not be able to pinpoint the cause of the trouble.[12] It seems that here Kuhn has been misled by his dealings with and understanding of Quine. Quine thinks that an adequate translation manual will not lead to such difficulties. "These examples," says Kuhn, "suggest that a translation manual inevitably embodies a theory, which offers the same sorts of reward, but also is prone to the same hazards as other theories. To me they also suggest that the class of translators includes both the historian of science and the scientist trying to communicate with a colleague who embraces a different theory."[13] At first sight it might appear that the sentences are talking about similar phenomena, and that the two references to "theory" ascribe to it similar roles. But on closer inspection this seems not to be the case. Translation manuals embody the theories and hypotheses that radical translators have about what the target speakers mean by their words. So the first mention of "theory" refers to a theory possessed by the translator. But in the second sentence the theory mentioned is one belonging not to the translator but to the target (the colleague, the historical scientist). And it is the latter that is of interest as regards the problems of translatability between theories from different paradigms, and is an aspect largely absent from Quine's thinking about translation (as Kuhn goes on to note).

On the whole the reference to Quine is neither especially helpful to Kuhn nor particularly germane.

Translatability, comparability and reference

We might ask, how important really is translatability for commensurability and comparability? So far we have assumed that the measure of a hypothesis is a set of relevant observation statements, and that the hypothesis is measured by a pairwise comparison of members of this set with the observational consequences of the hypothesis, seeing whether the pair are identical or not. Since we are dealing with the identity of statements, translatability into a common language is essential. But we might think that the identity is more than is needed: for an observation statement to show an observational consequence to be true (or false) it is not required that the one be identical to the other, or, correspondingly, to its negation. Nor, more significantly, is any such identity required for a qualified person to see that the consequence is verified or falsified by the observation. Similar remarks may be made about the comparison of theoretical statements for nearness to the truth; must the existence of relative verisimilitude require that the statements in question be couched either in the same terminology or in terminologies that may be translated into one another?

An important case where commensurability and comparability are available without translatability arises where there is a common reference even though the referring expressions may not be mutually translatable. If today someone says "The Prime Minister is a socialist" and another says "Tony Blair is not a socialist", then the two statements cannot both be true, and those who know the identity of the prime minister can see this to be the case. But "The Prime Minister" is no translation for "Tony Blair". Of course, as mentioned, to know that these are in conflict requires knowing that Tony Blair is the Prime Minister. But nothing in what Kuhn has said has ruled out the possibility of historians of science knowing relevant identities like this. The Babylonian name "Ishtar" was given to the planet Venus and was supposed to be a goddess. For this reason it may be unsatisfactory to translate "Ishtar" as "the planet Venus", but that is no obstacle to assessing the accuracy of Babylonian mathematical astronomy.[14] It is in part for this reason that issues of exact translation are typically of little interest to scientists – what they want to know is whether they are talking about the same things, stuffs, properties, and so on. To assess the truth of some piece of medieval alchemy or its verisimilitude relative to earlier or later beliefs, we need only know what "sal ammoniac" and "aqua regia" are (ammonium chloride, and a mixture of nitric and hydrochloric acids); we do not need to try to find terms that might be equivalent in sense ("ammoniacal salt" and "royal water" perhaps). Indeed, it is not clear what it is to translate substance terms like this, if it is not to give their modern English names. As the last phrase suggests, and as we shall see, substance terms are indeed *names*. While it may be useful to be told that "Akhenaten" is literally translated as "pleasing to Aten", of rather more importance to historians (and independent of that translation) is the knowledge that "Akhenaten" is another name for the king Amenophis IV.

Kuhn resists this defence against incommensurability by claiming that even reference does not stay constant through scientific change. The example he mentions more than once is that of the term "planet", which before Copernicus covered the Sun and Moon but excluded the Earth, while now it excludes the former but includes the latter. Thus, he says, the things "planet" was used to refer to has changed. This rejoinder is unconvincing for two reasons, but before seeing what these are it should be remarked that the term "reference" here is being used in a slightly strained way. For predicative expressions like "planet" philosophers tend to talk of the *extension* as being that set of things of which the expression holds: the extension of "planet" is the set of all planets, and similarly the extension of "gold" is the set of all gold things. In the case of gold we can also talk of the *reference* of "gold" which is something different from the extension; the reference is the *substance* gold, perhaps best thought of as the single entity, the natural kind gold. In the case of "planet" it seems as if there is nothing quite like this reference, unless it is the property of being a planet. Kuhn is here interested in what he takes to be the changing *extension* of the term. Now let us return to the problems with Kuhn's rejoinder. First, even if Kuhn is right in this case there is no reason to suppose that this shift has given rise to problems of incommensurability let alone non-comparability. He gives no examples of communication problems that have arisen from such a shift, nor does he argue that because of it Ptolemy's theory could not be compared with Copernicus' theory. Secondly, it is not even obvious that there has been a shift in extension. What seems to be right is that there has been a change in what people have *believed* to be the extension.

People once called the Sun a planet but no longer do so. It would be perfectly coherent to argue that the reason why is that they had a belief about the Sun that they no longer have (viz. the belief that the sun behaved like Mars, Venus etc. in having a large orbit about the centre of the local system). In this view, if the Sun is not in the extension of "planet" now, it never was. People can be mistaken about extensions just as they can be about references.

These remarks are not intended to demonstrate that Kuhn is wrong, just to show that he has not proven his case. It would be difficult to deny that radical shifts in extension have on occasion taken place so that we have in effect a new terminology, an expression with a new meaning. Kuhn makes a plausible (but not unarguable) case for this being the case with the term "element" after the chemical revolution, and perhaps "compound" and "mixture" as well. However, not every change in perceived extension is a change in actual extension and careful historical work has to be done in each case in order to decide the issue. As regards the term "planet" matters are not straightforward in this respect. It is certain that the term is not defined just by a list, since the discovery of new planets would then be ruled out. I think a better reconstruction of the concept of "planet" can be given that contradicts Kuhn's claim that the extension of "planet" or its Greek equivalent have changed. Observers have long been aware that certain heavenly bodies – Saturn, Jupiter, Mars, Venus and Mercury – had "wandering" paths different from those of the stars. They were thus classified together as the five stereotypical planets, with a presumption that they shared some nature that explains their different motions.[15] Ptolemy's treatment provided such an explanation – they are not among the "fixed" stars but all orbit the Earth in a manner determined by heavenly spheres and epicycles. In Ptolemy's system the Moon and Sun are in this respect just like the stereotypical planets, sharing the same kind of motion, and so they too were classified as planets. However, in Copernicus' explanation the Sun is quite unlike the other planets, in that it is at the centre of the system, whereas the Earth is like them, sharing their nature in orbiting around that centre. The Moon is a problematic case, since it shares some of the nature of the five stereotypes, in that it does go about the centre, but in other respects it is unlike them in having its primary focus of motion being the Earth. With the advance to Newtonian cosmology there is the added difficulty that fundamentally the explanations of the motions of the five stereotypes, and the Sun, Moon and Earth are all alike. They all move around their common centre of gravity under each other's

gravitational attraction. Nonetheless it is clear that there are significant quantitative differences in motion even if not qualitative ones.
The Sun is the overwhelming source of gravitational influence in the
solar system and the common centre of gravity is correspondingly
much closer to it. In the view I have just outlined, the concept of planet
is fixed by the explanation of the motions of the stereotypical planets.
Whatever is like them in this regard is also a planet. Thus unknown
celestial objects that share in this "nature" (Uranus, Neptune, Pluto)
are also and were always planets. Similarly a well-known object like
the Earth is and was a planet since it too shares this nature, even if we
did not always know that. By the same token the Sun was thought,
mistakenly, also to have this nature and was correspondingly
thought, again mistakenly, to belong to the extension of "planet".

If this account of the concept of "planet" is right then the extension
of that term has not changed. Kuhn has said nothing that supports
his claim that it has changed better than the alternative hypothesis
that it has not. The example of "planet" is not itself enough to show
that questions concerning translatability cannot be bypassed by
focusing on reference and extension. We could reach a very similar
conclusion by suggesting that "planet" is *defined* as "celestial body
whose primary motion is a large orbit about the centre of the system".
If this were right the extension remains fixed because the definition
remains fixed. Of course, such definitions are rarely given explicitly;
it might nonetheless be that this way of thinking of what it means to
be a planet is implicit in our usage of the term. If so we may call this
"definition" the *sense* or *intension* of the world "planet". The relationship between the extension of a term and its intension is simple: the
extension is the set of things of which the intension is true. The
extension of the term "planet" remains fixed as long as its intension
does.[16]

Among early logical positivists it was accepted that a term was
either an observational one, in which case it has its meaning given by
ostensive definition – its meaning is an associated perceptual experience or sensation. Or it is a non-observational term, in which case it
has an intension that in principle can be explicated, ultimately, in
terms of observational concepts. The rules that link the observational
and non-observational (usually taken to be theoretical) concepts are
meaning postulates. However, actual inspection of scientific theories
fails to identify specific meaning postulates as opposed to the postulates that give empirical content to the theory. And so later versions
of positivism regarded all postulates linking theoretical and observa-

tional concepts as being in part meaning postulates and partly empirical in content. Learning the meaning of theoretical terms is thus holistic; those meanings are not acquired one at a time but in groups of related concepts.[17] As we shall see Kuhn accepts something like this view of meaning where by "meaning" we mean sense or intension. What distinguishes Kuhn from the positivists is his denial of a fixed, foundational and theory-independent observational vocabulary that anchors meaning.

Kuhn's intensional shifts

Let us now consider the question of referential and extensional stability through scientific change. This is guaranteed if there is stability of intension. But if the intension changes then reference and extension may well be subject to change as well. In the case of "planet" there may be historical evidence that the intension of "planet" has changed since ancient and medieval times, refuting the alternative hypothesis I canvassed; in which case it would be possible that the extension has changed too, as Kuhn claims.[18] Given the background of a positivist or logical empiricist view of meaning, change in intension might come about as a result of two causes. First, if theoretical claims change then on the view discussed, because theoretical postulates play a part in determining meaning, intension will change too. Secondly, the empiricist view anchors the meanings of theoretical terms in the meanings of observation terms. Since in Kuhn's view the latter change as a result of a revolution, he should expect the former to change as well. Kuhn does not clearly distinguish the two grounds for change in theoretical intension; the lesson Kuhn draws is still clear, that we should expect intension and reference/extension to change when there is radical theory change.

Two examples illustrate Kuhn's view. The first concerns changing conceptions of a chemical compound in contrast to a mixture, in the late eighteenth and early nineteenth centuries. In the eighteenth century, dominated by affinity theory, a compound was regarded as something produced when two substances interacted producing heat, light, effervescence or something similar. A mixture could have its components separated by mechanical means. Furthermore, what might appear to be intermediate cases – solutions, alloys, mixtures of gases etc. – were also regarded as chemical compounds, because the affinity theory of chemical reaction seemed to be able to account for their properties, such as homogeneity. In the light, however, of the discovery that certain reactions took place only in fixed proportions, Joseph-Louis Proust regarded the latter as the mark of chemical combination. Therefore what his opponent Claude-Louis Berthollet held up as counter-examples to the law of constant proportions (combinations of substances in any proportion), Proust dismissed as mere mixtures and not compounds at all.[19]

Kuhn regards this as a shift of extension due to a shift in intension. The shift in intension is akin to a change in definition but is also more than that. The dispute could not be resolved by a simple agreement to use a wider range of terms, each defined in observational or operational terms, to capture the variety of different modes of combination – one term for Berthollet's mixtures defined in terms of mechanical separability, another for Proust's mixtures, defined in terms of not being subject to union in constant proportion, and so on. The problem, according to Kuhn, is deeper than could be resolved by "conventional conveniences". Rather, the difference reflects different theoretical commitments, to the theory of affinity and to the law of constant proportions respectively, or more generally to the "way[s] they viewed their whole field of research".[20] In consequence of this "the two men necessarily talked through each other, and their debate was entirely inconclusive".[21]

The second case concerns Newton's laws of motion and gravitation and our learning of the characteristic vocabulary employed therein. In *The Structure of Scientific Revolutions* Kuhn discusses the often made claim that Newton's laws can be regarded as a special case of Einstein's theory of special relativity, governing instances where v, the speed of objects under investigation relative to the frame of measurement, is much less than c, the speed of light. For example, in special relativity, the relationship between force, rest mass, and acceleration is given by: $F = ma/\sqrt{(1 - v^2/c^2)}$. When v is very small compared to c, $\sqrt{(1 - v^2/c^2)}$ is very close to 1, and the equation becomes $F = ma$, Newton's second law. If this was all there were to the relationship between Newtonian and relativistic mechanics that would suggest that the transition from the one to the other was not revolutionary in Kuhn's sense, since no drastic revision is required to the core claims of the former.[22] Kuhn complains that this oversimplifies matters. In particular it assumes that there is conceptual continuity between the two theories, and this Kuhn denies. The $F = ma$ derived from $F = ma/\sqrt{(1 - v^2/c^2)}$ is not Newton's law, it just looks like it. In these formulae we have employed variables representing mass, and implicitly space and time (via acceleration "*a*"), and also force, but these, says Kuhn, are *Einsteinian* space, time and mass:

> the physical referents of these Einsteinian concepts are by no means identical with those of the Newtonian concepts that bear the same name. (Newtonian mass is conserved; Einsteinian is convertible with energy. Only at low relative velocities may the two be measured in the same way, and even then they must not be conceived to be the same.)[23]

In a later paper Kuhn spells out in detail the nature of Newtonian concepts and the way that they are learned.[24] Among other points Kuhn emphasizes that such learning involves little that counts as a straightforward definition; instead knowledge of concepts is acquired through exposure to the actual use of the expressions as displayed in exemplary situations – the exemplars discussed in Chapter 3. Students learn not only the meanings of the expressions but also, at the very same time, the beliefs about the world contained in their community's paradigm. Kuhn tries to capture this combination of conceptual and empirical information in the phrase "stipulative description". Kuhn also says the examples of usage to which learners are exposed also employ other elements of the vocabulary they are trying to learn. Thus such concepts cannot be learned singly but only in clusters of related concepts. Thus both the holism and the mixed conceptual–empirical features of the positivist view of meaning are preserved in Kuhn's account.

Kuhn says that the Newtonian, quantified conception of force may be introduced using the spring balance. But this is no simple definition, since the understanding given thereby requires employment of what are generally thought of as laws of nature – Newton's third law and Hooke's law. More significant are the ways in which we may introduce the terms "mass" and "weight" as used in Newtonian mechanics. Kuhn envisages two routes. First, one may explain "mass" via Newton's second law ($F = ma$) and the various experiments designed to illustrate it. The law and the concept are acquired together. Thereafter, armed with the concepts of force and mass, one can discover that bodies exert a force on one another that is proportional to the product of their masses. Thus the law of gravitation is learned as a pure empirical regularity. This then provides the basis for the concept of weight, which can be thought of as a relational concept, the gravitational force between two bodies. The second route to these concepts starts also with the concept of force based on the spring balance but reverses the conceptual role of the law of gravitation and the second law. The concept of mass is learned via the former, again in a "stipulative description", and again the concept of weight can be learned as the gravitational force exerted upon a body. There are also other ways of learning the laws and concepts on Newtonian mechanics, for example by introducing weight via the spring balance and mass through its vibrations. Kuhn points out that in normal, non-revisionary science these routes all yield the same outcome, and that scientists, whichever route they have taken can be regarded as members of the same speech community. However, if anomalies occur that require changes to belief, then the route one has taken will make a difference. Those who have taken the first route regard the law of gravitation as a purely empirical claim and so will be able to countenance revisions to it, but not to the second law, which, says Kuhn, their language binds them to preserve. One the other hand, had they taken the second route a change to the law of gravitation would not be available; instead the second law would be subject to possible alteration.

Kuhn then considers anomalies that could not be resolved by either of these alterations, and so would require revision to both laws simultaneously. If so, one cannot avoid changing at least one stipulative description, something that plays a role in fixing the concepts being employed. The lexicon itself must undergo a change, brought about by:

> such devices as metaphorical extension, devices that alter the meanings of lexical items themselves. After such revision, say the transition to an Einsteinian vocabulary, one can write down strings of symbols that *look like* revised versions of the second law and the law of gravity. But the resemblance is deceptive, because some symbols in the new strings attach to nature differently from the corresponding symbols in the old, thus distinguishing between situations which, in the antecedently available vocabulary, were the same. They are the symbols for terms whose acquisition involved laws that have changed form with the change of theory: the differences between the old laws and the new are reflected by the terms acquired with them. Each of the resulting lexicons then gives access to its own set of possible worlds, and the two sets are disjoint. Translations involving terms introduced with the altered laws are impossible.[25]

To aid our discussion it is worth highlighting Kuhn's commitments:

(a) Key scientific terms have an intension that depends on certain theoretical claims.

(b) The dependence in (a) is *thick* – the intension depends on a wide range of theoretical claims, perhaps all those in the relevant theory. Note that we are not obliged to think every change to a theoretical belief results in a change in meaning. For example, if one takes the first of the two routes to understanding Newtonian mechanics, a change to the law of gravitation will not require a change in meaning. But the other side of this coin is that a change to the second law will lead to a change of meaning. Kuhn's dichotomy between the two methods of learning is schematic, and it might be thought that in reality our learning is really somewhere between the two, involving both. This would suggest that every theoretical claim plays *some* part in fixing meaning. In which case it looks as if every change to a theoretical belief might result in a change to meaning. Kuhn gives this impression when he parenthetically remarks, in support of these being different concepts, that Newtonian mass is conserved while Einsteinian mass is convertible with energy.[26] The Newtonian conservation of mass is independent of the laws of motion and gravitation. Thus if one had rigorously stuck to the first route (or the second) to learning the concept of "mass", mass conservation ought to be a purely empirical regularity. The fact that Kuhn thinks that differences over conservation reflect not merely empirical but also conceptual differences suggests that he thinks that in practice the determination of meaning is shared by all theoretical commitments. His intensionalism is very thick.

(c) The dependence in (a) is not only thick but also *strict*. That is, for some property to be the reference of "mass" that property must be truly described by all the relevant laws and other descriptions contained in the intension of "mass". A *loose* intensionalism might allow some of the descriptions to be false of mass. Similarly a strict intensionalism requires that an object in the extension of a term such as "compound" satisfy exactly all the descriptions contained in the intension. I shall later argue that at least as far as extensions are concerned, a loose intensionalism might be better suited to Kuhn's views than a strict intensionalism.

(d₁) As a result of a change in intensional meaning, there are changes in extension. Thus the intension of "compound" differs between Berthollet and Proust, and consequently there is a difference in extension, i.e. a difference in the sets of things correctly called "compound".

(d₂) It is implicit in Kuhn's discussion that although there is a shift in extension, neither the earlier extension nor the later extension are empty. For both Berthollet and Proust there are some things that are "compounds".

(e₁) Similarly, as a result of a change in intensional meaning, there are changes in reference. In the case of "mass", the change of meaning from Newton to Einstein means that the reference (or "referent" to use Kuhn's expression) of "mass" when used by Newton is not identical to the reference of the same word when used by Einstein.

(e₂) Again, Kuhn thinks that although reference changes, reference is successful both before and after the change.

The claims in (d₂) and (e₂) are those that express Kuhn's commitment to an incommensurability that is more than merely a change in intension.

Sparse and abundant properties

Points (d₁) and (e₁) raise an issue to which we will need to be sensitive, concerning the difference between extension and reference when dealing with property terms. By "property term" I mean an expression that names a property or predicates a property, quality, etc. of objects – and here I am using the term "property" in a loose sense with no especial metaphysical significance (yet). Thus "gold", "golden", "negatively charged", "compound", "mass", "weight", and so on are all property terms, and indeed most theoretical expressions are likely to count as property terms in this loose sense. Importantly these terms do not form a homogenous set with respect to their possession of extension and reference. As explained above, the extension of such a term is the set of things of which the term may be truly predicated, or which possess the property named. The reference is something distinct from the extension – it is the property itself or the natural kind conceived of as a quasi-abstract entity that transcends the extension. Thus the extension of "gold" is the set of all gold things, and the reference is the substance gold, the sort of thing we have in mind when we say "gold is denser than silver" or "gold has atomic number 57" and so on. Not all property terms are substance terms, "deciduous" for example, or "green". Both of these terms have extensions – the sets of deciduous trees and green objects respectively – but their references are clearly not substances. We can think of there being properties of being deciduous and of being green (additionally, in the latter case we might think that "the colour green" names an abstract object of sorts). Does every property term that has an extension also refer to some property (or substance, abstract object etc.)? Consider Goodman's term "grue". "Grue", for present purposes, is defined as attaching precisely to those objects that are green and observed before the year 2020, or blue and not observed before 2020. "Grue" has a perfectly well-defined extension which includes all emeralds currently in pieces of jewellery and those sapphires that will not be dug up until after 2020. But does "grue" refer to anything? Let us assume that "grue" is merely contrived and that grue things have nothing genuinely in common and no natural laws govern the behaviour of specifically grue things. We are inclined to think that in *some* sense of "property" there may be properties like being red, being deciduous, being negatively charged, and being a neutrino – there is no property of being grue. Properties in this sense we may call *sparse* properties. But we may choose to leave our metaphysical options open and think of there being a property for every property term with well-defined extension, including "grue", and these we will call *abundant* properties.[27]

Let us now think about the Proust–Berthollet controversy as Kuhn understood it. There is supposedly a shift in the intension and extension of "compound", and so for clarity we can talk of "B-compounds" and "P-compounds". The extensions of "B-compound" and "P-compound" are non-identical sets, although they do have a non-empty intersection. What properties are there here? Since there are two well-defined extensions there are two abundant properties. If modern chemistry is right, then there is a sparse property of being a P-compound, a substance formed by a union obeying the law of constant proportions. But there is no sparse property of being a B-compound. According to what is now believed "B-compound" is rather like a tame version of "grue", covering different kinds of stuff which have little genuinely in common, both certain sorts of mixture, like alloys, and substances formed by chemical reactions. The difference between "grue" and "B-compound" is just that the former obviously picks out no sparse property while it took a scientific discovery to see that there is no sparse property of being a B-compound. What then may we say about the difference in reference between "B-compound" and "P-compound"? If we think that reference can only be to sparse properties, then "B-compound" has no reference but "P-compound" may do. If we are happy with reference to abundant properties, then both refer, to different abundant properties.

Which view should we prefer? This may seem to be an abstruse metaphysical question, but since Kuhn talks of changing reference, it is reasonable to ask what is meant by this. When we explain facts we do so by referring to the properties of things: an object conducts electricity because it is gold; two objects repel one another because they are both negatively charged; a kind of tree is found in the valleys but not on the mountainsides because it is deciduous; and so on. Such properties appear in laws of nature and projectible empirical generalizations. These properties are the sparse ones. That something is grue explains nothing; similarly, being a B-compound explains nothing – if modern chemistry is right. If Berthollet had been right all along then perhaps being a P-compound would explain nothing, and there would be no corresponding sparse property, while there would be a sparse property of being a B-compound. The point is not who is correct, but that the properties we are interested in identifying and using for explanatory and other scientific purposes are sparse properties. Another way of looking at things is to consider that *abundant* properties are really only the reflections of our linguistic capacities. The abundant property of grueness seems to exist for no other reason than we can form the concept grue. But that is no more reason to think "grue" refers to a real entity in the world, than the fact that we can form the concept "the Queen's brother" is a reason to think that this phrase refers to a real human being.[28]

Thus, in the scientific context, we have reason to think that it is reference to sparse properties that is of interest. What are the consequences for Kuhn's views if we take it to be so? If reference is only to sparse properties we must deny that the term "compound" refers to one property when used by Berthollet and to another when used by Proust. Either reference is to the same property in both uses, or there is reference failure in at least one of the cases. Thus (e_2) cannot be true as regards the term "compound". We reach the same conclusion when we consider "mass". Let "N-mass" do duty for the term "mass" as used by Newtonians, while "E-mass" replaces the same word used by Einsteinians. Kuhn thinks that both N-mass and E-mass exist and are different from one another. Say, for sake of argument, that "E-mass" does have a reference that is not the reference of "N-mass" and also that it is E-mass that explains the behaviour of objects, not N-mass. In that case we would have no reason to think that N-mass exists as a sparse property. It is implausible that there are *two* similar but distinct properties that explain the kinematics and dynamics of objects.[29]

So let us consider instead what are the consequences for Kuhn of taking reference to be to abundant properties. This is in any case a more plausible interpretation of Kuhn's intensions. It allows (e_2) to be true, with reference to both of the distinct properties of being a B-compound and being a P-compound, and to both of N-mass and E-mass.[30] It reconciles Kuhn's view that there is reference with his scepticism concerning the truth of theories. To know that a property word refers to a sparse property typically requires knowing that the theory in question in true. So a sceptic about theoretical truth is not in a position to know that there is reference. But since reference to abundant properties depends only on the word's having a coherent application, that sort of reference can be known *a priori*.

Even if taking reference to be to abundant properties is a better interpretation of Kuhn there are a number of problems that make this an unattractive route. An initial difficulty is that the most obvious version of the abundant property view, that there is a property for every predicate that can be used in a true sentence, leads directly to a contradiction. For we can form the predicate "property that does not apply to itself", which thus defines an abundant property, P. It is readily seen that P applies to itself just in case it does not apply to itself – Russell's paradox. So even an abundant view of properties cannot afford to be too liberal and needs some kind of restriction.

Rather more importantly the abundant property view threatens to render vacuous Kuhn's claim of reference change. The abundant property view says that what part of what "(abundant) property" means is that if P and Q are predicates then P and Q refer to the same property if and only if P and Q have the same intension (subject to any further conditions designed to forestall Russell's paradox). It then follows immediately from the meaning of "property" that if a scientific property term changes its intension it changes its reference. In which case (e_1) although true is merely trivial, as is (e_2). The abundant property theorist can say that N-mass exists as an abundant property distinct from the abundant property of E-mass, but saying so does no more than rephrase the assumption that

"N-mass" and "E-mass" are different concepts. This leads to the third reason why the abundant property view is inappropriate in this context. As already mentioned, it is just an implausible way of thinking about properties and reference in the scientific context, for it allows reference to properties such as the property of being grue, just because we can define and use the predicate "grue" in a true sentence. But surely we want it to be a matter of scientific rather than linguistic discovery that certain properties exist and are available to be referred to. In particular it is sparse properties that appear in laws of nature and so make law statements true, false, near to the truth, and so on. And so it is continuity of reference to sparse properties that is relevant to the question of whether Einstein's equation represents a better approximation to the truth than Newton's. The claim that reference to abundant properties has changed is, therefore, not relevant to this discussion. For incommensurability due to reference shifting to be a problem for Old Rationalist views of progress it must be reference to sparse properties that we are talking about.

Intensionalism strict and thick

The last section considered the implausibility in Kuhn's (d_1) to (e_2). In this section I want to consider difficulties that arise in particular from adherence to (a) to (c). Consider the term "the Wallabies" as used in rugby union football circles. Plausibly the meaning of the term, its intension, is given by the description "the Australian national rugby union football team". Therefore someone who understands the term "the Wallabies" will in virtue of that understanding know that "the Wallabies are the Australian rugby team" is true. Such knowledge is *a priori*. In general, if "D" is a definite description that gives the intension of a term "T", then the statement "T is D" is knowable *a priori*. Furthermore, such statements are analytic – true in virtue of the meanings of the terms.

What has been said in the last paragraph is subject to a qualification. Imagine that Australia elected a government that outlawed the playing of rugby, perhaps led by a fanatical supporter of Aussie rules football. In which case Australia would have no national rugby team and the Wallabies (the team, not its members) would cease to exist. The term "the Wallabies" would cease to refer. In which case it is at least unclear whether any statement about the Wallabies could be true, including the statement "the Wallabies are the Australian rugby team". So what really is true (analytically and knowable *a priori*) is that if "the Wallabies" refers then the Wallabies are the Australian rugby team. Generalizing, where "D" gives the intension of "T", it is analytic and knowable *a priori* that if "T" refers, then "T is D" is true.

Let us now apply this to "N-mass", that is the term "mass" as used by Newtonians. Commitment (e_2) says that "N-mass" has a reference, i.e. N-mass exists. According to (a) and (b) the intension of "N-mass" is fixed by all the laws of Newtonian mechanics together. Of course, those laws mention "N-mass", but a formal device enables us to get an intension that does not employ the concept "N-mass". We remove every occurrence of the expression "N-mass" from Newton's laws and replace it with a variable, x. Now the laws taken together will form a complex predicate, which we may abbreviate as $L(x)$, so the intension of "N-mass" may be characterized as: "the property x such that $L(x)$". From what was said in the last paragraph it is knowable *a priori* that if "N-mass" has a reference (i.e. assumption (e_2) is true), then "L(N-mass)" is true. "L(N-mass)" is precisely the conjunction of Newton's laws. Hence Kuhn's account leads to the conclusion that it is knowable *a priori* that if his, Kuhn's, assumptions are true, then Newton's laws are true. Note that this conclusion is particularly unpalatable if one has an abundant conception of property. In that conception it is knowable *a priori* that "N-mass" refers, for that is guaranteed by its use as a well-defined predicate. Hence, by modus ponens, one should be able to know *a priori* that Newton's laws are true.

Kuhn seems to have a partial recognition of these consequences of his view, first calling laws "quasi-analytic" and later saying "I do not want to call such laws analytic, for experience with nature was essential to their initial formulations. Yet they do have something of the necessity that the label 'analytic' implies. Perhaps 'synthetic a priori' comes closer".[31] Changing the label does not really help eliminate the problems that are now apparent for an intensionalism that is both strict and thick.

(i) First, it is implausible that it is *a priori* that if "N-mass" refers then Newton's laws are true. If we are thinking of abundant reference then "N-mass" will refer whether or not Newton's laws are true. On the other hand if we are thinking about sparse reference then N-mass is just that sparse property, if there is one, Newtonians were talking about when they used the term "mass".

Might not they have been referring to the same property even though some slight variant on Newton's laws were true?

(ii) Kuhn's assumptions make the laws turn out true, whether analytic or not. Since Newton's laws are not all true, one at least of Kuhn's assumptions must be false. If we retain an intensionalism that is thick and strict, we must reject the assumption that "N-mass" refers – there just is no property of N-mass. But this is problematic for Kuhn. Imagine that Newton's theory and Einstein's both fail to be true. Then neither "N-mass" nor "E-mass" refers. In which case we cannot characterize the shift between one paradigm and the next as involving incommensurability due to a shift of reference.

(iii) More generally there is a problem for an intensionalism that is too thick. If all of the claims of the theory are involved in the intension of some key term, then if any part of the theory is mistaken two things follow: (a) the term has no extension; and (b) it does not refer to any sparse property. If the theory is at all sophisticated then the chances of some part not being true are high. So, for example, while much of Rutherford's theory of the atom was correct, some was not. But we would not want to say that the partial failure of the theory means that Rutherford failed to refer to atoms, electrons, protons and so on. On the contrary, he was talking about them, and he had some true beliefs about them and some false ones.

Thin intensionalism

The sort of intensionalism I have attributed to Kuhn is both *thick* and *strict*. It is thick because the intensions that Kuhn ascribes to theoretical terms include a wide range of conditions and descriptions – the whole of the content of a theory. It is strict since for something to be the reference of a theoretical expression it must satisfy *all* the descriptions contained in the intension (and be the only thing to do so). Kuhn could escape the conclusions of the previous section by making his intensionalism either less thick, or less strict. Less thickness would mean that fewer descriptions would be included in the intension – for example, not all of Newton's laws would be included. Retreating on strictness would mean that for something to be the reference of the term it need not satisfy all of the intension, only some sufficient proportion.[32]

What would follow were Kuhn to retreat on the "spreading out" of meaning discussed in assumption (b)? If not all laws are implicated in fixing meaning, then those that are not can be regarded as purely empirical and will not be true analytically or knowable *a priori*. Kuhn would have to give up the implication that the difference in conservation laws between Newtonianism and relativity show the references of "N-mass" and "E-mass" to be different since the conservation laws might be such purely empirical laws. He would have to accept that something like one of the two routes he describes is accurate. Some particular laws play an analytic role and others an empirical role. But if we retreat this far, note that it is possible to retreat yet further. Let us imagine that the second route were the one actually taken and it is the law of gravitation that fixes the concept of N-mass. Kuhn says that someone who had taken this route would not be willing to contemplate a revision to the law. Is this plausible? Strictly speaking the law of gravitation provides a specific value for G. When Newton introduced the law it was a law sketch, since he had no value for the constant G. This would seem to imply that Newton had no determinate concept of mass and that once a value had been supplied by Cavendish any subsequent change to that value was a change in the concept of mass. One way to avoid this is to break down the law of gravity into two components. The first says that gravitational force is proportional to the product of the masses and inversely proportional to the square of their separation. The second component says that the constant of proportionality has some particular value. We can now say that the first component supplies the concept of mass while the second is purely empirical. But we could go further and make the fact that the law is an inverse *square* law an empirical component. It could be that the analytic, conceptual component says simply that non-zero masses exert a force upon one another while the empirical components fill in the details. If we go this far it is no longer clear that there is any conceptual revision in the move to special relativity, since that leaves the Newtonian account of gravity untouched. Similar comments may be made about the first route. In *The Structure of Scientific Revolutions* Kuhn seems to have a very thick conception of intension. Later on the conception becomes rather thinner, and he says that it is the second route that gives us the concept of mass, remarking that the law of gravitation is fully contingent but that the second law, $F = ma$, is necessary.[33] But why is it that all of the content of the second law contributes to the intension of "mass"? We could say that what fixes the meaning of mass is the more general claim that mass is that property which explains an object's change in motion in response to an applied force. There is additional content required to get from this to $F = ma$, and that additional content would be purely empirical. If this were correct, then again there need be no change in concept in the shift to relativity, since inertial mass in relativity is also what relates force and change of motion.[34]

To summarize, Kuhn assumes that laws fix intensional meaning and thereby also fix, in a strict manner, extension and reference. If it is further assumed that all laws play a part – thick intensionalism – then it follows that if the term in question refers, then the laws must be true; correspondingly if one or more of the laws fails to be true then there is no reference, in which case incommensurability of reference cannot arise. But we do not need to make this further assumption of thickness: we can allow that some laws do not contribute to meaning. Furthermore, it need not be that the total content of a given law contributes to meaning, and indeed it is implausible that it should be so. But then the content of laws breaks up into an element that determines reference and an element that does not. It could be that the reference-determining element is sufficiently general as to be applicable to E-mass as well as N-mass, in which case those will be the same property. Thus there will be no incommensurability of reference in this case either. Consequently his argument fails to establish incommensurability due to shifts in reference.

Loose intensionalism

Strict intensionalism does not give Kuhn what he wants. Either way it makes incommensurability of reference unlikely: if too thick, because there is likelihood of reference failure, if too thin because reference may remain unchanged despite change of intension. What if Kuhn were to give up on the strictness in his intensionalism? In this section I will suggest that a *loose* intensionalism is better suited to Kuhn's aims, although it is not a view that Kuhn himself discusses. A loose intensionalism is still intensionalist – terms still have associated sets of descriptions, conditions, criteria etc. that play a role in fixing reference and extension. But now that fixing is not a matter of an entity satisfying the descriptions exactly. Concentrating on extension for the moment, we may say, for example, that an entity is in the term's extension if it satisfies X per cent of the descriptions in the intension. How high or low X is depends on how strict or loose the intension is (it might vary from term to term) – if $X = 100$, then we have strict intensionalism once again.

The advantage of this approach for Kuhn is that it blunts the first horn of the dilemma just outlined. Under strict intensionalism thickness leads to a danger of reference failure or empty extension since the defining theory is likely to be false overall and so provide an exactly true description of nothing. However, under loose intensionalism, even though the theory overall may be false, some parts of the theory may be true of some things, and so those things may enter the extension. To that extent it shares the advantages of thin intensionalism. But it also avoids what from Kuhn's point of view are its disadvantages, since loose but thick intensionalism also allows for both shifts in intension and shifts in extension. Shifts in intension are allowed for, since the intension is still the thick set of descriptions contained in the whole theory; so theory change leads to change in intension. And shifts in extension are also allowed for. This is because once the theory and so intension has changed some new entities might reach the X per cent threshold for membership of the extension, while other entities that once were part of the extension might drop out of it.

This might help treat the Proust–Berthollet debate in the way that Kuhn would like. "Compound" for Berthollet is fixed by affinity theory and related hypotheses while it is the law of constant proportions that provides the intension for Proust's use of the term. Affinity theory is false, and so in a strict intensionalist view, taking all of affinity theory to fix the intension of "B-compound" there are no B-compounds. Even so, not everything in affinity theory and its surrounding beliefs is false, in particular beliefs about the phenomenal nature of chemical reactions, such as the claim that heat is a characteristic product of reactions, or the belief that certain kinds of substance cannot be separated into components by mechanical means. So it may turn out that a number of substances are truly described by many, although necessarily not all, of Berthollet's theoretical beliefs. And so, according to loose intensionalism, the extension of "B-compound" is not empty. Furthermore, it may well turn out to include some substances we would call mixtures, since, for example, non-chemical dissolving typically does produce heat, albeit in small quantities, and some mixtures are difficult to separate with non-chemical techniques. If we move our attention to Proust's notion of a compound, we can see that it has a different intension from Berthollet's concept, since Proust employs a different theory. "P-compound" also has a non-empty extension. But it is not the same extension as the extension of "B-compound" since ordinary solutions and other such mixtures will not be described by a sufficient proportion of Proust's theory. In conclusion, the two concepts have different intensions. So they cannot translate one another.[35] And they have different although overlapping extensions. So there is no opportunity to ignore translation failure by concentrating on preservation of extension.

In describing how loose intensionalism can help Kuhn I have concentrated on extensions. What about reference to properties, as in the Newtonian and Einsteinian mass case? I think this is more difficult for Kuhn. To explain reference to individuals within a loose intensionalist framework, one might employ a "cluster" or "best fit" conception. The idea is that associated with a name is a cluster of descriptions; it might be sufficient for some entity to be the reference of that name than the object satisfy some of the descriptions, not necessarily all of them. Depending on how this idea is developed, it might be that the reference is that entity uniquely satisfying some particular proportion of the descriptions. Alternatively it might be that reference is to that entity that fits the descriptions better than any other entity. That allows us to be mistaken in some of our intension-fixing beliefs concerning an individual. A cluster view allows for shift of reference. It is claimed that among natives of the region, "Madagascar" once referred to the mainland of Africa, but since then because of European explorers and colonists the reference has shifted to the island. This could be explained by a shift in the cluster of beliefs attached to the name so that while the mainland made the majority of the natives' associated beliefs true, the island made most of the associated beliefs of the newcomers true. For such a shift in reference to be possible, there have to be two individuals in existence for the reference to shift between. This is what is implausible in the case of sparse properties. Say for sake of argument that Einstein's theory is right or nearly right. So "E-mass" refers to some sparse property that is for the most part described accurately by that theory. What about "N-mass"? For that to refer, there must be some sparse property that Newton's theory is mostly true about. But the falsity of Newton's theory and the superiority of Einstein's means that there is no reason to suppose that there exists any sparse property that is for the most part described truly by Newton – unless that property is the same property that is referred to by "E-mass". It looks then as if loose intensionalism is better at coping with Kuhn's claims about shifts in extension than with his claims concerning shifts in reference.

In suggesting that he might make his intensionalism loose as well as thick I intend only to show how Kuhn's views can avoid incoherence. Their truth depends on the truth of (d_1) and so of (b). Thick intensionalism is inherited from positivist views of theoretical meaning that we looked at in Chapter 1. As already mentioned in connection with the word "planet" it is prima facie at least plausible to think that there is a constant, quite thin intension that remains fixed. The extension is also fixed, although beliefs about the extension may change. Another case, not far from the concerns of Proust and Berthollet, is the meaning of the term "element". This term has had fairly explicit definitions that seem to have changed very little from Aristotle's time. Aristotle's own definition says that elements are substances "into which other bodies may be analyzed, which are present in them either actually or potentially, . . . and which cannot themselves be analyzed into constituents differing in kind"; Lavoisier held that elements are substances not analyzable by chemical means, while a modern textbook says similarly that an element is a substance that cannot be decomposed into simpler substances. While we should be careful not assume too quickly that Aristotle and Lavoisier meant entirely similar things by their definitions, neither should we conclude that they meant very different things just because they listed different things as elements. For it is perfectly possible for Aristotle and Lavoisier to have had roughly the same intensions but different beliefs about what things satisfied those intensions. Lavoisier himself listed certain substances as elements that we regard as compounds, for example magnesia and lime. But that does not show that he had a different conception of element from our own; it shows that he wrongly thought these to be undecomposable. Interestingly, Lavoisier did not include potash and soda as elements although no one had decomposed these. It is plausible that he was confident that these could be decomposed, thanks to an analogy with ammonia which Berthollet had decomposed. Sir Humphry Davy's development of electrolysis proved Lavoisier right about soda and potash but, later, wrong about magnesia and lime.[36]

The causal theory

So far I have considered whether Kuhn's case for incommensurability of both intension and reference can be made to stand up. I shall now turn to what other philosophers have said about the meaning of scientific terms in order to see what conclusions may be drawn for incommensurability. The most significant development involved a rejection of intensionalism altogether and its replacement by causal accounts of reference. In the face of difficulties with the causal view intensionalism has made a comeback, but the important conclusion to be drawn is that neither side of this debate provides any basis for a significant thesis of incommensurability.

The anti-intensionalist, causal theory of reference was born from a series of arguments attributable to Hilary Putnam and Saul Kripke.[37] Putnam argued that whatever determines the reference and extension of "water" it cannot be a certain sort of generalized scientific theory of the kind Kuhn has in mind. He asks us to conceive of a world called "Twin Earth" that is just like ours except that where we have H_2O they have another substance, whose formula we may abbreviate by "XYZ", which in all superficial respects is like water, appears in the places we expect to find water (rain, lakes, sea etc.) and supports life. The difference between H_2O and XYZ lies in their microstructures and requires some moderately advanced chemistry to detect. On Twin Earth they call XYZ by the name "water" (and presumably "eau", "Wasser", and so on as well). The word "water" as we on Earth use it includes H_2O but not XYZ in its extension. If a chemist created some XYZ in a laboratory we might be amazed by its water-like properties but we would not regard it as water. Now consider Twin-Earthians using their word "water" ("water$_{TE}$" for convenience). Does "water$_{TE}$" have the same extension as "water$_E$", our Earth word? If we say "yes", then a person on Twin Earth who points at some XYZ and says "That is water" is saying something false, since what they are pointing at is not water. But that is an odd conclusion. Why should our Twin Earth counterparts be accused of error when we are getting it right? Since the thought experiment is set up symmetrically between Earth and Twin Earth we must conclude that "water$_{TE}$" does not refer to water (it refers to the stuff XYZ). "Water$_{TE}$" and "water$_E$" have different extensions and different references.

Putnam now asks us to think about Earth and Twin Earth in 1750. The reference and extension of "water$_E$" have not changed in the last 250 years, and so neither has the reference or extension of "water$_{TE}$". And so in 1750 the references and extensions of "water$_{TE}$" and "water$_E$" differed just as now. If intensionalism were right then the two terms ought in 1750 to have had intensions that differ. Now consider what people knew about water in 1750. They did not have even moderately advanced chemistry and so they knew of water only things that are also true of XYZ – that it is tasteless, colourless, useful for quenching thirst and necessary for supporting life, but *not* that it is composed of hydrogen and oxygen. Hence they knew nothing that could be employed in an intension that would pick out water (i.e. H_2O) but not XYZ. Since we have already concluded that there must be a difference in intension between "water$_E$" and "water$_{TE}$" it follows that whatever these intensions are, they are unknown to the language users.

Note that we could reverse the argument, to show that if intensions are known then they do not determine extension. Either way, if Putnam's arguments are accepted, it follows that intensions cannot be what connects the user of a term to its extension or reference. So what does? The common answer among philosophers is that it is some causal link between the substance or object and the use of the term that forges reference. Such an account seems especially plausible with respect to proper names of people, things, places and so on. For example I know that both James Garfield and William McKinley were US presidents who were assassinated, but I know little else about them and nothing that would distinguish them. So I do not know any intension that serves to pick out Garfield but not McKinley. So what makes it that I am referring to Garfield when I ask someone "Was Garfield born in Ohio?" A sensible explanation is that if one traces the source of my usage of the name back through the occasions on which I heard this name and then through the sources of those usages and so on, the trail ultimately leads back to the man Garfield but not to McKinley. We can say something similar about our usage of the term "water$_E$". It is the stuff on Earth (H_2O) that plays a causal role in fixing our use of the term, while the different stuff on Twin Earth (XYZ) plays no such part. Hence "water$_E$" refers only to H_2O, and the extension of the term excludes the superficially water-like stuff on Twin Earth.

If this is the right way to think about reference then there would be no reason to think that theory change would lead to a change in reference. We can have different theories about what water is, and indeed some early hypotheses held that water is simply HO, without that implying any change in the reference of "water". On the contrary, what was being talked about remained constant and the different theorists were engaged in genuine disagreements, not in "talking past each other". The same could be said about mass. If there is a quantity property that we are causally acquainted with via its presence in instances of forces leading to changes in the motion of objects, then "mass" or "inertial mass" could name that property, and name it whether we are Newtonians or Einsteinians.

Putnam's argument works against both strict and loose intensionalism. Even loose intensionalism requires difference in theoretical beliefs for there to be difference in extension or reference, and it is just that difference in theory that is lacking in our comparison of Earth and Twin Earth in 1750. Kripke attacks loose intensionalism more directly, arguing that a speaker S could use a term to refer to some entity E even though S had very few true beliefs or many false beliefs about E. The latter means that S cannot have in mind an intension that E even largely satisfies; it might even be that some other entity F satisfies the supposed intension better than E. Taking the Garfield–McKinley example above, I do not have enough true beliefs about Garfield to constitute a cluster that he satisfies better than anyone else. Moreover it might be that I have some false beliefs about Garfield (e.g. that he was president at the time of the Spanish-American War), so there is a danger that were the reference of my use of "Garfield" to be determined by such a cluster, it might be that my use refers to someone other than Garfield – McKinley for instance. Kripke imagines that some of our most central beliefs about gold might be mistaken (such that it is yellowish in colour) due to strange optical illusions or other effects. A loose intensionalism could handle some of these beliefs being false, but not too many. To take a different case, it was long believed that whales were fish; no doubt people had or might have had many other false beliefs about whales, so that no cluster of such beliefs was better satisfied by whales than anything else (indeed such clusters might be better satisfied by some fish, such as the whale-shark). In all these cases it seems that reference is not achieved by an intension, even loosely conceived, but by a causal connection with the person, kind, or species being named.

It is not surprising therefore that Kuhn tries to resist the causal theory. Putnam imagines that a spaceship from Earth visits Twin Earth. The visitors might at first believe that their hosts mean by "water" what they themselves do, but eventually, having done some chemistry, they will report back that on Twin Earth the word "water" means XYZ. Kuhn responds to this by saying that what the visitors will report is that modern chemistry is all wrong, since according to our current beliefs there could not be a molecule that is as complex as XYZ and also evaporates at normal room temperatures and pressures.[38] A serious revision to our lexicon would be required. Kuhn has merely pointed out a tangential implausibility in Putnam's story; he muddies the waters (so to speak) but does not say why this deficiency undermines the point that Putnam is making. We can repair the story by imagining that the visitors knew enough chemistry to know that water is H_2O but not enough to have reason to believe that XYZ is impossible. Perhaps mid-nineteenth-century chemistry was in this state. But in any case, making a trip to Twin Earth and reporting back is not a key part of Putnam's story. The point he wanted to make in a graphic way was simply that the extension of "water$_E$" does not include anything composed of XYZ, which is why the space travellers would report that "water$_E$" and "water$_{TE}$" do not mean the same.

Kuhn's second point is directed at the identification of water with H_2O. It is tempting to draw metaphysical conclusions from Putnam's story, viz. that necessarily water is H_2O – in all possible worlds whatever is water is H_2O and vice versa. Indeed from similar stories also linked with the causal theory of reference, Saul Kripke draws exactly that conclusion with respect to gold and the element with atomic number 79. Kuhn complains that water is a liquid while H_2O can exist also as a solid and as a gas – ice and steam. So "H_2O" does not pick out only water. Kuhn relates this to changes in conception of substance in the 1780s when it became accepted that one chemical kind could exist in different states. Kuhn does allow that modern terminology can accommodate this, by talking of "liquid H_2O". But then, he says, we have used a combination of two properties to pick out the stuff we are interested in, which suggests that we could, in other cases, use a combination of several more properties. Would not this be a return to a descriptive, intensionalist account of kind naming? "Are we not back to the standard set of problems that causal theory was intended to resolve: which properties are essential, which accidental; which properties belong to a kind by definition, which are only contingent? Has the transition to a developed scientific vocabulary really help at all? I think it has not."[39]

Again Kuhn says things that are partly true but not to the point. The simplest way of seeing this is to appreciate that we do not have to draw the conclusion from Putnam's story that water and H_2O are identical. The only conclusion we are forced to is that in all possible worlds water consists (largely) of H_2O; nothing compels us to think the reverse, that every occurrence of H_2O is an instance of water. The important point in undermining the intensionalist account that Kuhn is associated with is that the stuff on Twin Earth (i.e. XYZ) just isn't water, whether or not competent speakers of English have enough theory or technique to distinguish the two. The issue of whether gaseous H_2O is water is simply irrelevant. It might be added that nothing in the causal view of reference says that we *cannot* ever pick out a kind by using a description. The causalist's claim ought to be that description, intension etc. cannot be all there is to reference or extension determination; sometimes at least reference is achieved by a causal connection that goes beyond what can be done with description alone.

Problems with the causal theory

Even if Kuhn's argument does not dent the causal account, there are nonetheless many problems with the simple causal story that suggest it needs some sort of supplementation. In the James Garfield case we may imagine that the causal history originates with a baptism or other naming ceremony. Similarly our learning about water involves, in our own past or in someone else's, a "dubbing" or naming of some sample of water. Even in the "mass" case we are acquainted with facts in which the property is instantiated. No doubt these facts explain why Kuhn is right in detecting that we learn what words mean by acquaintance with exemplary situations. However, not all theoretical terms could acquire their meanings in this way, since in cases such as "neutrino", "strangeness" (the name of a property of subatomic particles), and other names of unobservable entities and their properties, we are not acquainted with samples that can anchor our reference. Of course we can still be causally connected with such things, but then we are causally connected with all sorts of objects and property for which we have no name. What makes a causal connection a connection of reference in the absence of a sample with which to carry out a dubbing or baptism?

A related question concerns theoretical terms that fail to refer. Consider "phlogiston". There is no phlogiston, and so there can be no causal connection between phlogiston and our use of the term. In which case "phlogiston" fails to refer. That in itself is the conclusion we should expect. But two potential problems are raised. First, in this respect "phlogiston" is in the same boat as "caloric", "N-ray" and so on, all of which fail to refer. The simple causal theory makes no distinction between them, yet there is surely some sense of "meaning", close perhaps to intension, in which the meanings of these terms differ. And it is plausible to suggest that this meaning is typically understood by understanding the theories which surrounded these hypothetical entities and substances.

A second problem is that as described the simple causal theory is not very discriminating. There may be several things that play a key causal role in the use of an expression and the simple theory may be unable to choose between them or may choose the wrong one. If we think about phlogiston again, this substance was hypothesized in an explanation of combustion. What actually is involved in instances of combustion is oxygen. Hence there is a danger that because oxygen is causally responsible for combustion, which in turn is causally connected to our use of "phlogiston", it may be a consequence of the causal view that "phlogiston" is in fact a name for oxygen.

Causal descriptivism

On account of these difficulties the view has become widespread that the causal theory needs supplementation by a descriptive element, what we may call "causal descriptivism". In effect causal descriptivism seems to put us somewhere between the simple causal theory and the thick intensionalist view. Before progressing further we should recall that the causal theory was supposed to identify the reference of a term with the cause of our use of the term. In the case of unobservable and other inferred entities, the causal chain between entity and use is mediated by certain characteristic phenomena. Thus, in the case of "neutrino" the phenomena are various unexpected losses in mass in nuclear β-decay; for "hydrogen" it might be the production of water through combustion, for "(static) electricity" it might be various effects observable with comb and paper, and so on. Since these cause our use, what causes them will also be causes of our use. Therefore most discussions of the causal theory and its variants find it more convenient to talk in terms of the causes of these phenomena than in terms of the causes of our use.

The shift to talking about phenomena helps overcome the first problem for the causal theory, since we do not need to have an observable sample of the intended reference on which to perform a dubbing. Causal descriptivism seems to solve the remaining problem's difficulties. A causal descriptivist theory will regard the causal connection between t (the reference of "t") and the relevant phenomenon (and hence use of "t") as being supplemented by information describing, for instance, what sort of entity is being picked out, or the manner in which it causes the phenomenon. So the reference of "neutrino" is not fixed simply by its being causally connected with loss of mass during β-decay but in addition by its being described as an uncharged particle with a small mass that is emitted from the nucleus during β-decay. Now we can see why "phlogiston" fails to refer (as it should) rather than referring to oxygen (or perhaps deoxygenated air). For the supposed causal connection between phlogiston and combustion can be augmented by a description of it as a substance (implying homogeneity), a gas under normal conditions, that is present in combustible materials given off during combustion. Since none of the substances involved in combustion satisfy this description, "phlogiston" does not refer.

It might be noted that the causal element in causal descriptivism is sliding towards being an added description, one that asserts a causal connection. In the simple causal theory, a description like "water is causally responsible for our use of 'water'" would be true, and if we focus on the phenomena as the causal mediator between entity and use, "neutrinos are causally responsible for mass loss during β-decay" would also be true. The simple causal theory regarded belief in the truth of such descriptions as irrelevant to reference fixing; it is as sufficient for reference that the appropriate causal connection exist. But in causal descriptivism the supplementing descriptions must play their role by being explicitly or implicitly believed to be true of the intended reference. In which case it is difficult to see why the descriptions stating the causal connections should, or could, be excluded. If the supplementing descriptions I associate with "neutrino" include "low mass neutral particle emitted during β-decay" it is difficult to see how I could fail to have in mind that they are causally responsible for a loss of mass during β-decay. Thus causal descriptivism perhaps should not be described as a beefed-up causal theory – a theory asserting a causal connection plus descriptions. Rather it is a pure descriptive theory, with the feature that some of the descriptions must state that the entity that is the intended reference plays a causal role in some specified phenomenon. As Howard Sankey says of his own version of causal descriptivism, "the reference of theoretical terms is determined by description of the causal mechanism, whereby the action of unobservable referents is thought to produce certain independently specified (e.g. by ostension) observable phenomena".[40]

This sort of descriptivism can be reached via a route that starts much closer to Kuhn. David Lewis tells us how to define theoretical terms thus:[41] We take the theory, T, in which the theoretical term in question, "*t*", is found. We then take out every occurrence of "*t*" and replace it by a variable term, say "*x*", giving us T(x), which can be thought of as one long predicate. The proposition $\exists!x$ T(x) says that there is exactly one entity (object, property, kind etc. as appropriate) such that T(x) is a true of description of it. Lewis's idea (inherited from Ramsey and Carnap) is that we can define *t* as just that entity, the unique entity of which T(x) is true – in formal terms, $t =_{\text{def}} \iota x$ T(x) (where "ι" is the definite description operator – "the x such that . . .").

Lewis recognizes, as does Kuhn, that T will likely contain more than one theoretical term, and so the explanation just given will have to be generalized. More importantly, this approach concurs with Kuhn's view that the meanings of theoretical terms are defined by the theories in which they are found. What I want to show is that despite the fact that the Lewis view can capture the last mentioned intuition behind Kuhn's account of meaning, it in no way endorses his view of radical failure of reference sharing across theory change.

Before doing so it is worth mentioning that the Lewis view is not in conflict with causal descriptivism as I have characterized it. As David Papineau, who promotes the Lewis view, points out, terms like "cause", "physically necessitates" etc. will have to remain in the T(x); they cannot be treated as theoretical terms. Hence it is perfectly possible for the definition of "*t*" to include some description of the causal role of *t*s; it might start: "$t =_{\text{def}}$ the x such that xs are causally responsible for phenomena $P_1, \ldots P_n$, and xs are particles . . ." and so on. Thus the T in the Lewis account might be precisely the set of descriptions employed by the causal descriptivist, including a description about the causal role of the intended reference.

If the Lewis view captures Kuhn's idea that scientific terms are defined by the theory of which they are a part, why does Lewis's view not lead to Kuhn's conclusions about reference? The first matter to deal with is Kuhn's claim that a term may refer both before and after a theory change yet have different references. I have already argued that this is not consistent with Kuhn's own view that the way a theory fixes the reference of a term is both thick and strict, and the Lewis account helps make the point. If "*t*" is defined thickly and strictly by the whole of a theory T then "*t*" refers only if T is true – or, to be exact, if and only if there is a unique entity satisfying T(x). If we have to change our theory because T has been proven to be mistaken, then there is no reference for "*t*" – *t*s do not exist. Of course, we might be erring in changing our theory and it was true all along. But to the extent that the replacement theory, T*, is supposed to remedy a perceived defect in the earlier theory, T and T* cannot both be true, and since T is true T* will be not be. In which case the term "*t*" in T* will not refer, even if "*t*" in T does refer. Indeed both theories might be false in which case "*t*" fails to refer in both contexts.

The only occasion on which there will be genuine shift of reference will be where a theory and its "replacement" are both true (but about different things). But in such cases it is not clear why the "replacement" is genuinely a replacement. That is not to say that cases like this may not exist. From various causes, including simple misunderstanding, the use of a term may shift or become transferred. No doubt this may cause difficulties for historians, translators and interpreters. But there is no reason to suppose that this is a pervasive feature of theory change, and indeed, given that it involves transference from one true theory to another one might suppose it *not* to be a feature of revisionary theory change.[42]

Hence when T is replaced by an incompatible T*, "*t*" as defined first by the one then by the other will at most refer in only one context, so the idea of a Kuhnian shift of reference is ruled out. If a shift of reference is ruled out, what should we say about shift in intension, and so about untranslatability and incommensurability? The Lewis account is an account of term *definition*, not merely reference fixing, and so shifts in intension that follow shifts in theory are not excluded. T might be replaced by a compatible T*, where both theories determine the same reference. T might say that water is a substance that is a liquid under normal conditions, that is necessary for life on Earth, that freezes at 0°C and boils at 100°C; T* might say that water is H_2O. Here we have two different intensions picking out the same substance. There is continuity of reference through change of intension. If intension is what must be preserved in translation, "*t*" or "water" as used in T* does *not* translate "*t*" ("water") as used in T. Sankey says of his causal descriptivist account: "My argument for translation failure assumes that the way in which the reference of a term is determined is a semantic property which must be preserved in translation".[43] Here the semantic property, the *way* in which reference is determined, is the term's intension. Sankey's is explicitly an account where translation failure is accepted but denied to be a problem for theory comparison, since the latter requires only continuity of reference. The verdict is that incommensurability (as translation failure) exists, but is no big deal.

Thin intensionalism revisited

The last discussed cases of shifting intension concern the replacement of one defining theory by another where the two are not in conflict. The supposed change for "water" was the replacement of a somewhat superficial theory by a compatible but more sophisticated one. These are not the cases that are of interest to Kuhn, where there is revisionary theory change. The problem of revisionary change is that one of the theories must be false, so the term defined by the false theory must fail to refer. That may be fine for "phlogiston", but as we have discussed, this seems to mean that "mass", "gravity" and so on in Newtonian mechanics all fail to refer, and similarly for other now discarded theories. Indeed it may be a problem for current theories, since these may be false too. An advantage of the simple causal theory, that now seems to be lost, was that terms in false theories could refer and maintain their reference through theory change. Continued reference was supposed to be the anchor for theory comparison. And note that we want to be able to compare theories that might both be false – one can still be better than the other.

The most direct solution to both problems is one that simply bypasses the whole issue of shared intension *and* shared reference. Returning to the explanation of the Lewis account, the empirical content of the theory T is carried by the proposition $\exists!x\ T(x)$.[44] Therefore when comparing T and T* all we need do is compare $\exists!x\ T(x)$ and $\exists!x\ T^*(x)$. Since the theoretical terms have been removed from these, the issues of meaning continuity and reference continuity of theoretical terms is simply irrelevant.

Papineau offers a more nuanced solution that runs in parallel to the discussion I gave of Kuhn's view earlier. The solution is to adopt a more thin intensionalism than hitherto. The question we should ask is: how much theory is there in the theory T that defines "*t*"? The immediate point to accept is that not *every* belief about *t*s has to be included in T. Consider that we might be in a situation where we have already defined "*t*" by reference to T and then believe we have discovered something new about *t*s, $P(t)$. $T + P(t)$ now sums up our total theoretical belief about *t*s, but only T defines "*t*". Hence we could discard and replace $P(t)$ with $P^*(t)$ without changing the meaning or reference of "*t*". So how much theory is needed? Untranslatability, incommensurability and problems with Kuhn's position arose precisely because he wanted much, perhaps all, of what we believe to be included in the defining theory – he had too thick a conception of intension. Including too much in T(x) promotes the danger that nothing satisfies T(x). Including all the axioms of Newtonian mechanics in T will mean that "mass", "gravity" and so on will not refer. But at the same time we should not include too little in T(x). The term "t" defined by T refers only if there is an *unique* entity satisfying T(x). If T(x) is too thin, then several distinct things may satisfy T(x), and so on this account "t" will not refer. There is also the danger that if T(x) is too thin it might be satisfied by a unique entity but the wrong one because the intended, intuitive referent of "t" does not exist. This problem is exemplified by concern that with too little defining theory ("x is a substance the exchange of which explains combustion") "phlogiston" might even refer to oxygen.

Nonetheless, between the scylla of too much theory (and so false) and the charybdis of too little theory (and so underdetermining reference), there is a reasonable amount of room for manoeuvre. For many common theoretical entities the amount of theory required to determine them is quite small: "component of air that is responsible for combustion" uniquely picks out oxygen, "cell whose function is the transmission of electrical signals" determines reference for "neuron". Given that many other things are true of oxygen and neurons respectively (oxygen is the element with atomic number 8, neurons are normally polarized, etc.) we should expect there to be many sets of propositions that uniquely determine the appropriate references. Therefore we can say that it is not especially difficult for the following situation to exist: a set of true beliefs, T_{core}, determines the reference of "t". T_{core} is supplemented by additional beliefs, T_{supp}, about ts that do not contribute to the meaning of "t". Our total theory of ts, $T_{core} + T_{supp}$, can change over time, because T_{supp} changes, while the reference of "t" remains constant, since T_{core} remains unchanged. As discussed above, as soon as Kuhn admits that there can be beliefs that do not contribute to meaning (i.e. beliefs in T_{supp}), he has no reason (or has given no reason) to suppose that the situation just described is not commonplace, even when there is radical theory change. For T_{core} can be quite small and T_{supp} correspondingly large, and so revision of all of T_{supp} is consistent with continued reference to ts by "t".[45]

The reason for discussing the Ramsey–Lewis–Papineau approach at length is that among the worked-out positions on the meanings of scientific terms the Papineau view is the one that comes closest to accepting Kuhn's assumptions (a) and (b), by agreeing with Kuhn that such terms have intensions and that these intensions are fixed strictly by the part a term plays within a theory. Despite this close-ness to Kuhn we do not have to accept his views about change of refer-ence, and consequently we do not have to concur with him about the difficulties allegedly created for theory comparison.

Taxonomic incommensurability

We have discussed reference and theory-dependent intension at length for two reasons: first, in order to see what exactly Kuhn thinks on these important topics – and to show that what he does think, that intension depends thickly on theories, is rather closer to positivist views of meaning than to contemporary ones; secondly, in order to examine Kuhn's response to one defence against claims of incommen-surability. That defence maintained that even if a shift of intension did take place, theory comparison was not jeopardized since that required only continuity of reference. Kuhn denies that reference does stay the same, but as we have seen it is very difficult to maintain both that reference is successful before and after a shift in intension and that the reference has changed. In particular Kuhn does not succeed in showing that this has actually happened in important scientific cases.

The foregoing rests on the assumption that the entities, kinds and properties that are available for reference are standing features of the world. If there is Einsteinian mass, then that was there for New-ton to refer to or not; if there is no phlogiston, Priestley could not have referred to it even though he tried to and thought he did. In the light of the world-change thesis discussed in the Chapter 4, might Kuhn be unwilling to accept this assumption? As we saw, the world-change thesis is most plausible as a thesis about the nature of experience, not one about theoretical entities. To extend the thesis to the latter would be to make it an implausible version of social constructivism.

Kuhn's world-change thesis may nonetheless be employed in support of incommensurability to the extent that the latter is limited to our empirical vocabulary. If our view of the observable world changes then so may our use of the words we use to describe it change also. For example we might imagine that our colour words divide up the visible spectrum in a different way, as linguists tell us other languages do. Here reference to sparse properties seem less relevant than the shifting of extension, and I have pointed in this direction already, that shifts in extension are more easy to accommodate than shifts in reference. In this section and the next two I shall look at Kuhn's view that incommensurability amounts to translation failure due to changes in extension that result from the re-organization of our classificatory vocabulary.

Kuhn developed his account of incommensurability as untranslatability in the 1980s and 1990s. A necessary (but not sufficient) condition on translation is preservation of *taxonomy*. Taxonomies serve to divide up the world. A taxonomy is created by a *lexicon*, which is a set of terms that are holistically related to one another, forming a *lexical network*.[46] The holism of these terms is local, involving a relatively small group of expressions (such as "mass" and "force" in Newtonian mechanics), rather than spreading to include all of a language's vocabulary. Changes in the meaning of one of these terms cannot be achieved without changes to the meanings of others. Kuhn talks of these taxonomies as attaching to the world thanks to criteria of application. Revolutions involve changes in taxonomies. The criteria of application change, and because of the holism the criteria change for all terms in a taxonomy. This shift results in changes to the extensions of the terms. One metaphor that goes back to *Structure* is of a taxonomy as a grid that we lay down on the world to measure or describe it. Taxonomies, like a grid for map references, are, says Kuhn, conventional. A shift in taxonomy is like picking up the grid and putting it down again in a different orientation. (The holism of a taxonomy means we have to pick up the grid as one rather than moving just part of the grid.)

Certainly things like this can happen. We could imagine that "north" was once defined conventionally as the direction of the magnetic pole, but then was redefined as the direction of the true, geographic pole (where the axis of the Earth's rotation meets the planet's surface). As a result of such a change the utterance "*X* is north of *Y*" might change from having stated something true to something false. The extension of the predicate "north of Watford" would be different before and after the change. It seems fair to say that "north" would have changed its meaning; holistically, "south", "east" and "west" would have changed their meanings too. Kuhn's claim is that this happens to the classificatory terminology of a scientific field when it undergoes a revolutionary development, and that its occurrence prevents translation between the language used before the change and the language used after the change. In Kuhn's view a lexical term has, in general, non-empty extension before and after revolutionary change, *and* the extension changes. We have just examined the difficulties faced in maintaining this view – unless Kuhn adopts a loose intensionalism. I suspect that Kuhn's later views on taxonomy are best understood as operating within a loose intensionalist framework, even though he never makes this explicit.

In examining how Kuhn's idea of taxonomic incommensurability is developed in detail we should bear in mind four questions. The first question is: why does a shift in taxonomy prevent translation? Might there not be a rule for moving between taxonomies? Secondly, even if there is no direct rule, might there not be a taxonomy that translates terms of both the taxonomies, thus allowing for indirect translation? Thirdly, why cannot we just use both taxonomies (being careful to avoid ambiguity) in one language? Then there would be no question of translation, since a sentence expressed using one lexicon would automatically be expressed in the language of which the other lexicon is a part (even if that lexicon is not itself used). The fourth question is: why should we worry about translation? So what if translation in Kuhn's sense is impossible? The example or metaphor of the map referencing grid suggests that fairly simple answers that favour the possibility of translation may be given to the first three questions. Given a map reference using one definition of north, a formula will translate it into a map reference based on the other definition. It is true that no simple translation of "X is north$_{(magnetic)}$ of Y" is available in terms of "north$_{(true)}$", but a perfectly exact translation is available that makes use of the formula mentioned. The change from imperial to metric measures is like this, and furthermore the terminologies of both systems are part of modern English. And so even if there is no perfectly exact formula of conversion (employing finite decimal notation), as would arise if metres and yards are mathematically incommensurable, it would still be false that "x is 10 metres long" has no translation into English, since it already *is* English, and likewise for "y is 20 feet long".

Sankey says that Kuhn takes the locally holistic nature of the meanings of scientific terms to be a cause of untranslatability between shifted taxonomies.[47] Although Kuhn is not entirely explicit on the point, Sankey's interpretative claim is plausible. One might argue that Kuhn is right on the grounds that if terms "U" and "V" are part of a local holism, then the sense of "U" will be captured in part by its relation to "V". If we attempt to define "U" in terms of some vocabulary, L, that does not include "V", then the true sense of "U" cannot be captured, for the link with "V" is not also captured. There are two reasons why this is a weak basis for the Kuhnian claim. First, definitions of both "U" and "V" in terms of L can perfectly well preserve the precise nature of the relation between them and so preserve sense, and, indeed, doing so will be a necessary condition for the successful definitions of the terms. Imagine that there were a slight shifting in the application of our vocabulary of general colour names ("red", "green" etc.). We might nonetheless possess a vocabulary for finely discriminated shades ("vermilion", "crimson", "olive", "puce" etc.) so that the general terms of both the original and the shifted lexicons could have exact definitions in terms of these finer shades. The second reason for rejecting the argument for untranslatability from holism is the fact that Kuhn allows that a taxonomy may be preserved even though the criteria for the application of its terms have changed. He acknowledges that new discoveries might allow for new ways of applying the terms but without changing the extensions of the terms. It is only revolutionary discoveries leading to alterations in extension that are problematic.[48] It might be, therefore, that non-revolutionary discoveries are made that allow for an extension-preserving, taxonomy-preserving redefinition of the lexicon. If holism does not prevent extension-preserving redefinition of the lexicon, why should it prevent extension-shifting redefinitions?

Thus holism does not make untranslatability inevitable. What other reasons are there that might make us expect translation to be impossible? One reason why the terms of shifted taxonomies do not have common translations or a translation formula is that, in certain cases at least, the terms of a taxonomy are learned through the acquisition of similarity relations from exemplars. As we have seen Kuhn thinks of these in primarily perceptual terms. For example, Kuhn discusses a father teaching his son the meanings of the terms "duck", "goose" and "swan" by pointing to instances of these creatures while out for a walk in the park.[49] Having being told which are which the child acquires a sense for certain similarities – and dissimilarities – that allow him to be able to recognize ducks, geese and swans. Since the dissimilarities in particular are learned by reference to the different kinds ("that's a goose, not a swan"), there is a local holism; the three kind terms are learned together. Because this is not learning by verbal definition, if the child has not previously learned a vocabulary employing these exemplars, then the child will have learned a lexicon which has no verbal equivalents among expressions he already knows. In dividing up the waterfowl into discrete kinds, this lexicon provides a taxonomy. Kuhn does not properly explicate what he means by the "criteria" that link a taxonomy and the world, but in this case we may well guess what he would say. The criteria here may well not be verbally articulable differences, but that does not mean that the child is not sensitive to genuine differences. The length of neck is likely to be a criterion distinguishing ducks and geese; it is a feature to which consciously or unconsciously the child's perceptual apparatus has become sensitive.

We can use Kuhn's example to illustrate the sort of incommensurability he has in mind. Imagine that we shifted the whole taxonomy, so that some duck-like geese came to be included under the revised term "duck", and with similar shifts at the existing borderlines between "duck" and "swan", and between "swan" and "goose". If this new taxonomy is learned in a similar way, by reference to exemplars (some of which are now exemplifying the use of a different term, and will thus forge new criteria), then the new taxonomy will not be verbally definable in the child's old vocabulary. A fortiori, there will not be any language that will be usable for the purpose or translating or inter-defining the terms of the new waterfowl taxonomy and the old.

This will answer the first two questions posed above, why there is no rule for translation and why there is no third taxonomy available to translate both taxonomies. But note that it does so only because the taxonomies are created by similarity relations learned from exemplars. Kuhn thinks that this provides a model for all scientific learning, and so all scientific taxonomic language will be like this. As I have emphasized in this and previous chapters the exemplar idea is too simplistic to be much more than a metaphor when it comes to rather more theoretical science and the terminology it employs. To describe the immediate constituents of the atom we have a lexicon of "electron", "proton" and "neutron". Nothing like seeing exemplary ducks is involved in learning the structure of this taxonomy. As Kuhn himself acknowledges, such concepts are acquired through knowledge of a theory. He does emphasize the role of experiments and other forms of perceptual experience along with theory in the case of the Newtonian concepts. Let it be that a taxonomy rests on an exemplar that has a theoretical component and a perceptual one. Might not a revolutionary change affect the theoretical part without changing the role of the perceptual component? For example, Kuhn describes the role of experience in helping learn the basic concepts of Newtonian mechanics. But these experiences are neither replaced nor used differently in explaining Einsteinian concepts. Indeed what is somewhat ironic about Kuhn's repeated emphasis on the shift from

Newtonian to relativistic physics in illustrating incommensurability is that the way that most people get to learn relativity theory is by first getting to know Newtonian mechanics and then by learning how relativity differs from it. If incommensurability were a *serious* problem this would appear to be a very *bad* paedagogical strategy. But more to the point, the transition is not made by having any new experiences but purely by understanding some new theory. So the story about seeing ducks is not gong to tell much about the nature of this taxonomic shift (if there is one). Furthermore, one can learn a lot of theory and a lot of taxonomy without making any experimental observations or having any experiences of the relevant subject matter at all. The subatomic taxonomy is a case in point. Learning it is almost always achieved purely by understanding theory. Insofar as any laboratory experiences are involved they are usually only for the purpose of coming to believe the theory, not for understanding it or its lexicon. One might argue that a shift in taxonomy for empirical concepts may ramify through into a corresponding shift for theoretical terms. But this is by no means obvious, even if one adopts the empiricist assumption that theoretical expressions have their meanings rooted in the meanings of observational expressions.

The lack of a translation rule and of a common vocabulary for translation are to be expected only if we restrict our attention to perceptually based taxonomies. It is worth asking how much that really is? It might be tempting to remind Kuhn of the theory-dependence of observational vocabulary, but Kuhn's own thesis is really the *exemplar*-dependence of perception. The question is how much observational vocabulary *in science* is dependent on perceptual exemplars as opposed to verbalized theory? Relatively little, I suggest. Even the child's taxonomy of waterfowl is not the same as, or at least not established in the same way as the zoologist's, especially if the latter is employing a taxonomy based on genotype as opposed to phenotype. Kuhn's supposedly shifting taxon of "planet" is connected with perceptual exemplars only in the most attenuated way. Spotting a planet at night requires either prior knowledge or patient observation to see whether it changes position relative to the fixed stars. A neophyte cannot learn the concept (either the new or the old version) just on the basis of *seeing* a similarity between celestial objects picked out in the night sky; some inferential knowledge or theory is required as well.

Such cases also help us see another limitation of Kuhn's emphasis on the perceptual examples. In the child's taxonomy if something doesn't look like a duck it isn't a duck. Scientific taxonomies are not like that – appearances can be deceptive. For a zoologist something might look just like a duck but yet be a goose if there are good theoretical reasons for classifying it thus, just as we do not classify whales and dolphins as fish. Perceptually learned similarity relations are helpful but not the last word. Kuhn might well describe such rejections of apparent similarity as shifts of paradigm. But it is at least equally plausible to suppose that insofar as a taxonomy is scientific it tends to suppose that underlying a superficial similarity there is a deeper unity *and* that the taxonomy attaches to the latter not to the former. A scientific taxonomy of mammals requires no stretching or shifting to include dolphins, even if we previously thought them not to be mammals.[50]

To make the present, taxonomic, account of incommensurability work, Kuhn needs to make an implausible restriction to perceptually acquired taxonomies.[51] Even then he has not, for all we have discussed so far, excluded an evasion of the problems of translation that simply adopts both taxonomies within a single language. This might lead to dangers of ambiguity, but they could be overcome by appropriate tagging of the terms.[52] The fact that we can be bilingual in – and so understand – both of two languages (such as English and French) that have mutually untranslatable components, suggests that there is no obstacle to understanding a language with combined taxonomies.[53] So what prevents its existence?

The no-overlap principle

Kuhn's answer to this question is given in one of his very last papers, published in 1991. Here Kuhn introduces the *no-overlap principle*: "no kind terms ... may overlap in their referents unless they are related as species to genus".[54] The no-overlap principle is a restriction on the permissible structure of a taxonomy. It presents a bar to translation and communication thus:

> Shared taxonomic categories, at least in an area under discussion, are prerequisite to unproblematic communication, including the communication required for the evaluation of truth claims. If different speech communities have taxonomies that differ in some local area, then members of one of them can (and occasionally will) make statements that, though fully meaningful within that speech community, cannot in principle be articulated

by members of the other. To bridge the gap between communities would require adding to one lexicon a kind-term that overlaps, shares a referent, with one that is already in place. It is that situation which the no-overlap principle precludes.[55]

Sankey uses Kuhn's example of celestial taxonomies to illustrate Kuhn's view:

[T]o translate the Ptolemaic term 'planet' into the Copernican lexicon would require incorporation into the latter taxonomy of a single category containing members of three distinct Copernican categories. [viz. 'planet', 'sun', 'moon'] But no such category may be introduced as a natural kind of the Copernican taxonomy, since the Ptolemaic category combines entities together as members of a single kind which the Copernican scheme treats as members of distinct natural kinds.[56]

Kuhn's views on incommensurability of taxonomies are elaborated by Ian Hacking.[57] Hacking thinks of a taxonomy as a class of what he calls "entities" (and I shall call "types", so as to distinguish them clearly from the things – particulars – which will be instances of these types; the sort of type we will mainly be interested in are *kinds*). Taxonomies are structured in a certain way by a relation K, which is supposed, typically, to be the *kind of* relation. K is transitive and asymmetric.[58] So, for instance, if tarantulas are a kind of spider, and spiders are a kind of invertebrate, then tarantulas are a kind of invertebrate, but spiders are not a kind of tarantula. Taxonomies have a head type – a type to which everything belongs which if K is the "kind of" relation, Hacking calls a *category*. So the following illustrates a taxonomy with the category "invertebrate":

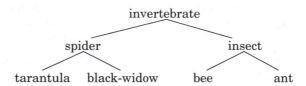

Hacking contrasts taxonomies such as the above with what he calls *anti-taxonomies*:

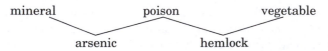

Here the types overlap – some poisons are also minerals, and some are also vegetables. Hacking excludes such overlapping in taxonomic classes, thus accommodating Kuhn's no-overlap principle. Some taxonomies have *infima* types. A type T is an infima type if there is no lower type which is a type of T. Say we have a taxonomy with the category "dog", with "Chihuahua" as a type of dog, our taxonomy might include no types of Chihuahua; in which case "Chihuahua" is an infima type.

Hacking uses this machinery to express what he takes to be Kuhn's view of natural, or, as Hacking prefers "scientific" kinds. There are three conditions on Kuhnian natural/scientific kinds:

(1) scientific kinds are taxonomic;
(2) taxonomies of scientific kinds have infima kinds;
(3) scientific kind terms are *projectible*.

A term is projectible if we use it to make inductive inferences – projections. Since "tarantula" is part of our scientific taxonomy, we may use that term in making predictions: "if that is a tarantula, it will have a poisonous bite". But some types are non-scientific and are not projectible; it is difficult to think of inductive predictions that one would make on the basis of something's being an *artwork*. Or we may invent a term meaning "person with a name whose second letter is 'o'"; again we don't expect these people are have something else in common, something dependent on their having a name whose second letter is "o", and so we won't use such a term as the basis of inductive inferences.

Why should a paradigm shift involving a change of taxonomies of scientific kinds lead to translation failure between those kinds? Hacking consider three cases:

(a) There is an overlap between a new kind and an old kind. Since by condition (1) scientific kinds are taxonomic, they obey the no-overlap principle. Hence the old kind cannot be a kind in the new taxonomy.
(b) An old kind subdivides a new kind which has no sub-kinds in the new taxonomy. The new kind is an infima kind, so no term in the new taxonomy translates the old kind term.
(c) Although the new and old taxonomies as a whole are not the same, the old kind and new one do coincide. In this case, Hacking says, we should not argue for untranslatability.

Does the untranslatability identified here really exist, in cases (a) and (b)? Regarding (b), rather than removing kind divisions, we are more used to scientific developments providing *new* divisions of kinds, as occurred when the existence of different isotopes of the same element were discovered. But Kuhn is thinking of revolutionary change, so matters may be abnormal. Perhaps the erstwhile kind of witches is a subdivision of the kind "human" for which there is no modern counterpart. Presumably phrenology had divisions of bumps on the head into kinds where we today recognize no kinds. The issue is once again whether these old kind terms are introduced by verbalized theory or by perceptual examples. If the former, then presumably we might be able to find theoretical explications of the concepts. Insofar as there was a science of witches, presumably that science had some theoretical principle whereby people who were witches were differentiated from those who were not (e.g. "people who were in communion with the devil"), and phrenology similarly may have divided cranial protrusions either theoretically, or by location. Without more being said, there is no reason to think that the principles of subdivision could not be explained in vocabulary that is independent of either taxonomy, i.e. in terms of some shared language. (It would not necessarily be an observational language.) In particular, it need not be especially complicated in this case, since ex hypothesi, the old kind subdivides a new kind, and so the old kind may be specifiable in terms of the new kind (which may also correspond to some old kind) plus some additional condition (e.g. witch = a kind of person who . . .). Regarding (a) similar remarks may be made. Hacking may have shown that an old kind term may not be translated by a new kind term. But that shows only a lack of direct mutual translatability between elements of the taxonomies, not a failure of translatability of an old kind term into the new taxonomy *plus* the rest of language.

Remaining with (a), we must ask why the no-overlap principle should be respected. If we drop it there should be no obstacle to mixing old taxonomic vocabulary with the new; nor, more generally, is there anything to prevent there being a language that employs both old and new taxonomic classifications. In fact perfectly unified, coherent sciences and branches of sciences employ taxonomies that violate the no-overlap principle. Chemistry supplies many such cases. For instance, we may classify elements by their group or according to whether they are metals or non-metal. Indeed the classification into metal and non-metal cuts across the taxonomy of pieces of stuff as belonging to particular elements, since some elements, such as tin, may exist in a metal form or as the non-metallic allotrope. Similarly the division of isotopes as radioactive or stable cuts across their classification into elements. In organic chemistry a variety of classifications may be used which cross-cut, such as regarding alcohols as a kind and phenols as a kind – kind that share some members in common. In biology two specimens may be classed as belonging to the same genotypical kind but different morphological kinds or vice versa. In none of these cases does the use of one classification preclude simultaneous use of another that cross-cuts it.[59]

Kuhn seems to think that one reason for respecting the no-overlap principle, and thus for not mixing taxonomies, is that doing so will permit conflicting projections. (In which case, Hacking's condition (3) is not independent of condition (1).) Let us say that "water" has undergone a taxonomic shift. We avoid ambiguity by employing the terms "water$_1$" and "water$_2$". If these violate the no-overlap principle then something might be both water$_1$ and water$_2$. But, Kuhn says, the two terms induce different expectations. Employing both of them is an unstable state of affairs to be resolved only by recourse to matters of fact. "And if the matters of fact are taken seriously, then in the long run only one of the two terms can survive within any single language community."[60]

This complaint doesn't exclude overlaps in general. The cases cited of actually overlapping taxonomies show this. Although the differing classifications of the same stuff will, given what is believed, permit differing expectations, those expectations need not conflict. The same stuff can have many properties according to which it is classified, and each of these may involve it in some law. But a thing may be subject to more than one law, like an electron which is subject to gravity, electrostatics and the laws of motion. So projectibility is no reason to exclude every case of overlapping taxonomy.

The case that is relevant for Kuhn's comment must be that where taxonomy P is introduced by a theory T_P and Q by a theory T_Q, where T_P and T_Q are incompatible theories, for example where one has replaced the other. Under strict, thick intensionalism one of the taxonomies will have empty types – so no problem there, since there is no overlap. Under strict, thin intensionalism the types (their extensions) may be the same for the two theories despite their incompatibility. Would this not give rise to incompatible projections, as a result of using T_P and T_Q respectively? While that would generate some incompatible projections, those projections would not be legitimated *merely* by use the taxonomies. What generates a taxonomy and fixes the extensions of its types is not the whole theory but some (usually) true core, which in this case will be common to T_P and T_Q.[61] The only projections authorized merely by membership of the taxonomic type will be ones generated by the true common core, and so will be the same whether one is using taxonomy P or taxonomy Q. The total theories T_P and T_Q will generate conflicting projections, but only on the basis of the taxonomies *plus* some additional theory, which differs in the two cases.

What then about thick but loose intensionalism, the view that seemed most congenial to Kuhn's view of overlapping extensions? Matters are rather less clear here. Let a taxonomic type R be generated by a theory T_R. According to loose intensionalism any object x will fall within R if it satisfies X per cent of the descriptions contained in T_R. Are we entitled on the basis of the classification of some object **a** as being in R to make every projection generated by the theory T_R? At first it might seem so; after all, the very meaning (the intension) of "R" is created by T_R; indeed no other theory could have given rise to that intension. So the use of the term "R" seems to involve a commitment to T_R, and thus to the projections generated by R. On the other hand, there is no *logical* commitment to every projection generated by T_R just on the basis of accepting "**a** is R". Let "x is F" be one of the theoretical descriptions associated by the theory T_R with being R. "x is F" is thus part of the intension of "R". But one is *not* logically entitled to infer from "**a** is R" that "**a** is F". Why not? Remember that according to loose intensionalism it is sufficient for **a** to be R that **a** satisfies only X per cent of the descriptions associated with R. And so **a** could be R without satisfying the description "**a** is F". For this reason, while users of the two taxonomies might be disposed to make conflicting predictions, it does not seem that their advocacy requires them to.[62]

Does untranslatability matter?

Let us conclude with my fourth question: does untranslatability matter? Above I suggested that two statements U and V might belong to mutually untranslatable languages yet a suitably qualified person who understood both languages might know that a certain fact, perhaps a fact known observationally, confirms one and refutes the other. One case we have discussed at great length is that where the key components of "U" and "V" have different intensions but the same references. Another case that shows that translatability does not matter for the philosophy of science concerns the sort of untranslatability more recently discussed, that due to shifting extensions. Names of shades of colours in different languages can be particularly difficult to translate. But that does not prevent us from making many cross-linguistic inferences. Imagine a language, English*, which is like English but where the corresponding colour terms cover slightly different regions of the spectrum. In English*, red* covers most things that are red, but includes some things that are reddish orange. Green* excludes some shades of green but includes various shades of turquoise and greenish-blue. Thus "red" does not translate "red*" and so on. But if one theory, couched in English*, has the consequence that some object *x* is red* while another theory, in English, says that *x* is green, then it is clear to someone who understands English and English* that the two theories are in conflict, and such a person may be able to make an observation to decide between them.

The thought that translatability is not important is reinforced by emphasizing what a very *weak* notion untranslatability is. Let *t* be a term of the language L which has no synonyms in L. Let L* be L minus the term *t*. Let it also be the case that *t* has no non-circular analysis into necessary and sufficient conditions expressible in L. Then *t* has no translation in L* and so L is taxonomically incommensurable with respect to L*. Logical empiricism held that the number of such terms *t* is quite small, limited to an important class of expressions, typically held to be ostensively defined, which provide the stock of basic expressions of a language in terms of which all the others can be given proper analytic definitions. But in the last 50 years it has become ever more clear that terms like *t* are the rule not the exception. Wittgenstein drew our attention to the existence of family resemblance concepts, which have no analytic definition, while the sort of causal and causal-descriptive accounts discussed above imply that theoretical and kind terms also lack analytic definitions. So the existence of taxonomic incommensurability looks like a significant claim only against a logical empiricist background. In the current case the terms of a theoretical taxonomy will be exactly the sort of term we should expect to be incommensurable with the remainder of the language of which they are a part. Consequently, where competing theories introduce new and different lexicons we should expect the terms of one theory's taxonomy to be incommensurable with the terms of the taxonomy of a competing theory. Some theoretical expressions may well be given explicit definitions. But since the definitions will be in terms of other members of the theory's taxonomy, such terms will still be taxonomically incommensurable with the terms of a competing theory.

Conclusion

The thesis of incommensurability is the claim that there is no common measure for theories from distinct paradigms. In this chapter we have largely been discussing the thesis in the form that there is no common language for expressing the competing theories or into which they may be translated. A particular cause of this is the fact that key terms change their meanings as a result of scientific revolutions.

A summary of our conclusions is best presented as they relate to different classes of term. First, we may consider perceptual expressions. That these should change their meanings as a result of revolutionary theory change follows from Kuhn's thesis of the theory-dependence of observation. But as we saw in the last chapter, the basis for his version of the thesis is weak and certainly does not extend to revolutionary changes of a highly theoretical sort. Perceptual experience may change if one's perceptual history changes in a significant way but the effect of theoretical change alone is marginal. Consequently there is no particular reason to think that, in general, revolutions bring with them changes in the meanings of perceptual vocabulary, although it is quite possible that a revolution concerning paradigms with a highly perceptual element could lead to incommensurability of this sort.

A second class of expressions is the class of non-perceptual, non-scientific expressions. For obvious reasons these were not given any special treatment. But they are worth mentioning here because they represent the most plausible case for a thesis concerning meaning change. Here the idea is that the extension of certain terms may change over time. Suggestive is the synchronic case of translation between different languages. Kuhn points out that the simple "the cat sat on the mat" has no straightforward translation into French because the word "mat" has no exact French counterpart.[63] Words like "tapis", "carpette", and so on get fairly close but have extensions different from "mat". Could we not imagine the synchronic difference between French and English becoming a diachronic difference between two periods in the history of a single language? We could

conceive of cultural changes leading to shifts of extension corresponding to the difference between "mat" and "tapis". The suggestion then
is that a cultural change leading to a shift in the extension of non-scientific words has a parallel in a scientific revolution, leading to a shift
in the extension of scientific terms.

This leads to the third case, theoretical terms of science. Is the
parallel with non-scientific, "culturally-dependent" words valid? No
it is not, the reason being that scientific terms are intended to play
roles that are different from those of non-scientific terms. The first
difference is that scientific terms are used in the framing of scientific
explanations, in particular explanations that refer to laws of nature.
To do so they need to pick out standing, general features of the world
– sparse properties. Most uses of non-scientific terms are purely
descriptive as in "the cat sat on the mat". There is no general property of "mattiness" that explains the common behaviour of all mats
and which distinguishes them from non-mats. (Compare this with
"metal".) A change in the conventions and criteria surrounding the
application of words such as "mat" does not undermine its ability to
be used in descriptions. A change in the use of "metal", so that the
word now covers plastics but not metals less dense than water, would
still permit the use of the word in descriptions –"iron is a metal"
would still be true. But we could no longer explain or predict the behaviour of a thing (e.g. its being an electrical conductor) by reference
to its being a metal. That is not to say that there cannot be explanations that appeal to non-scientific properties. That something is an
artwork may well explain why it is valued, why people make an effort
to see it, and so on. But the features of the world that such explanations depend on, being social or cultural, are variable. This is the
second difference between scientific terms and non-scientific terms
whose meanings are culturally dependent.

The use, design, construction and cultural significance of floor
coverings varies from place to place and so we would expect words in
different languages to be used to describe different kinds of such
thing. Similarly cultures change over time and we should not expect
words like "work of art", "scientist", and so on to have the same extensions in different periods of history. The criteria for the application of
such terms change because those features of the world they are used
to describe are changing. The scientific case is clearly different, since
the features of the world that scientists aim to talk about are fixed and
unchanging. Many non-scientific terms are applied with regard to criteria whose presence may be more or less manifest. Wittgenstein's

"game" is an example of this and it is significant that Kuhn's example of the geese, swans and ducks is like this too. But this is *not* characteristic of scientific terms. The latter are not applied on the basis of potentially variable criteria. As the causal, causal-descriptive and the Ramsey-Lewis–Papineau accounts of theoretical meaning all concurred, it is the causal explanatory nature of a term's reference that makes it the reference. (What makes oxygen the reference of "oxygen" is that it is the substance that *explains* the combustion of materials.) This means that the semantics of these terms is such that, inevitably, if they refer at all they refer to uniform kinds and properties that are fixed features of the world and which may be used in explanations.

This sort of semantics assumes a certain minimal realism: that there is a fixed world of explanatory properties and kinds out there to be referred to. What if this minimal realism is wrong? What if some strong sort of world change thesis were true? If minimal realism were false that would not show that this account of the semantics of scientific terms is false; at most it would mean only that, contrary to appearances, terms such as "oxygen", "neutrino", "mass" etc. failed to refer. In any case there is little reason to suppose that minimal realism is false. Certainly Kuhn's world change thesis is not strong enough to refute minimal realism nor is it intended to.

Kuhn's break with logical empiricism went only so far. He accepted the empiricist view that scientific terms get their meanings by some sort of implicit definition from the bulk (or even all) of the theory in which they play a part. This (thick) intensionalist viewpoint takes it that "sophisticated" concepts (like scientific ones) get their meanings via some sort of function of other, less sophisticated ones. Kuhn's only real difference with empiricism on this score is that he denies that basic "observational" concepts are fixed by reference to perceptual experiences that all human being have. That denial helps generate Kuhn's more radical conclusions about incommensurability and (some cases of) non-comparability, but only in concert with those empiricist assumptions he keeps. Had he jettisoned those, then some of the problems would not have arisen (shifting of reference and extension through revolutionary change) and others would have seemed commonplace (inability to define theoretical expressions in terms independent of the relevant theory). The causal theory of reference was devised not to combat Kuhn's views in particular, although it was readily put to that use; instead the intended contrast was the intensionalism of logical empiricism in general. Looking back over the last 40 years the causal theory was the radically new departure in semantics and consequently Kuhn's views, for all their radical-sounding conclusions, seem positively conservative.

The philosophical concern with incommensurability originates with Kuhn's desire to undermine old rationalist conceptions of progress.[64] If "mass" meant something different in the Newtonian and Einsteinian paradigms, then despite the fact that the equations of the former *look* like limiting cases of the equations of the latter, we cannot straightforwardly say that this is indeed the case, in which case we may not infer that the change in the equations represents an extension of knowledge or a better approximation to the truth. However, Kuhn does not establish the antecedent of the claim. It is not obvious that the intension of "mass" changes between Newton and Einstein and it is even less plausible that the reference changes. And it is worth adding that even if Kuhn were right about meaning change, that would not show the traditional views to be wrong – only that they are not so obviously right. For the conceptual shift involved might itself help contribute to increased knowledge and nearness to the truth. In the next chapter we shall see what further arguments Kuhn brings to bear on the traditional conceptions of progress.

Chapter 6

Progress and relativism

Kuhn's conception of scientific progress

The simple cumulative view of science – science as the discovering of truths and scientific progress as the growth of a stockpile of true scientific beliefs – is clearly mistaken. Many past scientific beliefs have been found to be in error and most likely many still are. However, while individual theories may have false consequences, and so are false overall, they may nonetheless have important true consequences. So a more sophisticated view of progress takes both truth and falsity into account. Propositions and theories may be nearer to, or further from, the truth. Progress is made when an old theory is replaced by new one that has greater *verisimilitude* – truth-likeness.

Kuhn rejects this account of scientific progress. His own views have been caricatured as saying that progress during normal science is relative only to the governing paradigm and that changes of paradigm are irrational leaps from one way of doing science to another. In fact Kuhn has a far more solid notion of progress than this, one that does allow for improvement across paradigms, and he rejects the charge of relativism.[1] He replaces the notion of progress as increasing verisimilitude with a conception of science as growing in problem solving capacity. Progress during *normal* science can easily be understood as continued success in solving problems. Something similar may be said about progress *through revolutions*. Since revolutions are revisionary, we cannot simply say that problems continue to be solved, albeit in a different paradigm, since that in itself would indeed be consistent with starting all over again, with no continuity with what went before. Typically there is continuity despite a revolutionary change. For one thing, the cause of a revolution is the existence of certain troubling unsolved puzzles in the earlier paradigm. A condition on a new paradigm being acceptable is that it solves or resolves at least some of these anomalies. For another, an acceptable replacement paradigm should "promise to preserve a relatively large part of the concrete problem solving ability that has accrued to science through its predecessors".[2]

In this chapter I shall assess Kuhn's account of progress and the challenge it poses to Old Rationalist accounts. As we shall see this raises a range of important topics: truth, scientific realism and relativism. In support of the main theme of this book I aim to show that Kuhn's challenge employs arguments that are to be found in the doctrines of the logical empiricists. In this case Kuhn's epistemological assumptions are thoroughly Cartesian, despite his repudiation of particular elements of the Cartesian tradition. Although the conception of scientific progress as increasing nearness to the truth is a conception that was held by the logical empiricists, it is important to note that this doctrine is not definitive of that movement. The opinions held by positivists and other logical empiricists (for example the list of views about observation mentioned on p. 97 above) that differentiate them from other philosophical movements neither entail nor are entailed by this account of progress. Indeed there is something of a tension between logical empiricism and the claim that science shows increasing verisimilitude. To the extent, which was considerable, that the problem of induction was central to logical empiricist concerns they had difficulty in showing that science actually satisfies their account of progress. In particular, Popper's critics accused him of being unable, in his own terms, to show that scientific change led to increasing verisimilitude even though he said it did. Furthermore, the thought that science is progressing towards truth is typically understood in a realist fashion: science is getting nearer to a mind-independent truth; thus understood, the claim of increasing verisimilitude is in tension with the anti-realism that is typical of positivism.

Hence there is no contradiction in saying that Kuhn rejected progress-as-increasing-verisimilitude for reasons he shared with positivists and other logical empiricists. In the second half of the chapter I shall sketch an epistemology that rejects those shared reasons. *Naturalized* epistemology is more optimistic about the possibility of absolute knowledge and correspondingly is potentially more friendly to the thoughts (a) that we can know that some scientific theories are closer to the truth than their predecessors and (b) that scientific knowledge is growing overall.

On Kuhn's behalf I shall also suggest that naturalized epistemology is nonetheless in tune with some important themes in *Structure*, themes that suggest that Kuhn could have made a more fundamental break with empiricism. Because we will be dealing with some basic issues in epistemology the discussion in this chapter will from time to time leave Kuhn's explicit arguments behind and the views discussed are sometimes more Kuhnian than Kuhn. This is inevitable if we are to discuss the ramifications of unstated assumptions and if we are to see Kuhn in a larger context, especially as he relates to recent developments in the philosophy of science.

The evolutionary analogy

Kuhn's theory of paradigms and the associated discussion of scientific education aim to explain why science progresses – in Kuhn's sense of "progress". Our understanding of scientific development is complete without any need for a view of science as getting increasingly close to the truth. A fruitful way of looking at Kuhn's sense of the superfluousness of truth is to understand scientific progress in evolutionary terms. While the evolutionary history of a species shows unmistakable progress of some kind, it is wrong to think of this as progress *toward* something. Convergence on the truth may be likened to pre-Darwinian notions of evolution that are understood teleologically – species strive to achieve some perfect form, perhaps as laid down by God. But natural selection showed that there are no ideals towards which species are progressing. Think of a hunting animal such as the cheetah. It has clearly evolved to run fast, since faster running cheetahs are more likely to catch gazelle and so be able survive and propagate. The fact that gazelles are fast runners too is explained similarly – the faster runners are more likely to escape and breed. Notice that this means that there is no optimal high speed which cheetahs are evolving towards. Were there such a speed then once gazelles had evolved to be faster, cheetahs would be in a tight spot, facing starvation. At that point there would be selective pressure on cheetahs to evolve to be yet faster than this supposedly optimal speed. In principle both species should evolve to be ever quicker.[3]

The evolutionary analogy is certainly fruitful when considering the development of a science. (It is perhaps both ironic and suggestive that the other philosopher to emphasize the analogy with evolution through natural selection was Kuhn's long-standing adversary Karl Popper.) The proliferation of scientific fields bears a resemblance to speciation, while Kuhn's picture of normal science interrupted by revolutions might be compared to the model of punctuated equilibria. Nonetheless, we might ask how close and helpful the analogy is for Kuhn's purpose of denying that theories get closer to the truth. Consider the evolution of giraffes. There is no advantage to a giraffe in being much taller than any tree. If the height of trees is independently fixed, then there is an optimum height toward but not beyond which giraffes will evolve. The lack of a goal in the first case considered was due to the fact that the two species were evolving in competition with one another. In a sense, at any given time the two species are evolving towards goals, but because of their interaction the goals are continually shifting. Our question is then, which is a more accurate analogy for scientific development: the more complex scenario with two evolving species interacting with each other, or the simpler scenario with one species evolving against a fixed environment?

The basic events driving evolution include attempts by cheetahs to catch gazelles. If a cheetah succeeds that increases the chance of her surviving and breeding, if she fails, the chances are reduced. The parallel in science is the testing of a theory against evidence. If a theory makes a prediction that is confirmed by the results of an experiment, then we are more likely to keep that theory, add to it, use it in conjunction with other theories and so on. If, on the other hand, the experimental results conflict with the theory, then it is more likely that we will reject or revise the theory. The difference between the two-evolving-species scenario and the one-evolving-species scenario is a difference in the historical nature of the "tests" the cheetahs undergo. In the one-species scenario the test remains constant over time – the trees eaten by the giraffes remain the same height – but in the two-species scenario the difficulty of the tests changes – the gazelles get faster over time; furthermore, the gazelles are getting faster *because* the cheetahs are improving in speed. In this crucial respect the single-species scenario is a better model for science. The results of experimental tests do not change. A good experiment is one that is replicable; it gives the same results whenever performed. (It is true that some sorts of measurement change, when we are measuring a variable quantity like the time of high tide.

But we think that the underlying laws remain fixed. Experimental tests do not change in a way that makes it more difficult for a theory to pass them, and even less do they do so *because* the theory is developing.) If a hypothesis passes a well-designed and well-executed experiment we are usually confident that it will continue to pass tests of that kind if repeated, and that any extension of the theory will thus also pass the test. In broad terms, the single-species model captures the idea that in science our theories may change but the features of the world that they respond to are what they are independently of our theories, and are by and large constant over time. Recall that in the single-species model we *could* say that the giraffe is evolving towards an ideal and that there would be no pressure on the species to evolve once that ideal had been reached. Similarly we can say of an area of scientific research that it will evolve towards an ideal beyond which it cannot go. In the limit the giraffe reaches an optimal "fit" with its (fixed) environment; a theory can reach a optimal fit with the world, and this would be a true representation of it, since only true theories cannot be falsified.[4]

To my claim that the bits of world that science seeks to represent do not change as a result of science, it may be replied that in Hoyningen-Huene's interpretation of Kuhn the phenomenal world *does* change as a consequence of a scientific revolution. So perhaps the case, from this perspective at least, is more like the two-species scenario. But there is crucial difference. The phenomenal world changes as a result of revolution in a direction that makes the fit between expectation and experience *more* close and *more* likely. Old exemplars are abandoned because they generate anomalies – unfulfilled or refuted expectations. As we have just seen, Kuhn thinks that revolutionary progress involves the resolution of such anomalies. So in terms of a changing phenomenal world, the new world will not generate a conflict with expectation where the old one did. If anything, the changes in the phenomenal world induced by a new theory are changes *towards* that theory. The opposite is the case in the two-species scenario. The gazelles do change in response to the increasing speed of the cheetahs, but they change not by making it easier for cheetahs to catch them but by making it more difficult. The world changes *away* from the cheetahs.

Kuhn thinks that the notions of truth and nearness to the truth are unnecessary in the explanation of scientific development. But even if he is right that does not rule out the possibility of our showing that in the long run scientific beliefs do in fact get closer to the truth.

The mechanisms that Kuhn describes might not have truth as their goal, but they may nonetheless be well suited to producing truth. Truth or increasing nearness to it might be a by-product of scientific activity. The cheetah does not aim at increasing the speed of her species, only at catching her supper; even so, the increasing speed of cheetahs is what will come of many cheetahs chasing their prey. Of course it may not be coincidental that most scientists, unlike cheetahs, will *say* that they aim at discovering the truth.

Epistemological neutralism

We have seen that Kuhn adopts a stance of neutrality on epistemological questions, at least when the latter are understood in a strong, absolutist way. The explanation of the history of science has no need to decide whether a particular puzzle-solution or shift in theory counts as an addition to knowledge. We do not need to refer to the truth of a hypothesis in order to explain why it was accepted. This sort of neutrality is explicitly endorsed as the touchstone of a certain kind of relativism, the Strong Programme in the sociology of scientific knowledge. According to David Bloor and Barry Barnes, the progenitors of the Strong Programme, sociologists and others seeking explanations in science should treat alike beliefs that we might judge to be true and those we think are false, and similarly those we judge to be rational and those we think are irrational.[5] This is the *equivalence* or *symmetry postulate*. This contrasts with positivist and falsificationist accounts as well as traditional history of science that concentrated on successful scientific heroes. According to such accounts (to a caricature of them at least), rational scientists will employ a methodology that is likely to lead to true theories. Successful science needs no more explanation than this. But by the same token, when scientists make mistakes and back false theories, some special sort of explanation is required, for example that some feature of the scientist's psychology or political pressure caused them to believe what they did. The symmetry postulate denies that we should make this distinction. Since it is plausible that on occasion at least we can explain scientific beliefs with regard to psychology, sociology or politics, the symmetry postulate seems to say that we should always provide explanations of that sort. In the wake of *The Structure of Scientific Revolutions* many historians and sociologists of science have been keen on referring to extra-scientific causes of belief such as political preference. Kuhn himself has very little to say on such matters, although he is of course sensitive to the general cultural background. His own historical work, for instance his masterly survey of early developments in quantum theory *Black-body Theory and the Quantum Discontinuity 1894–1912*, refers overwhelmingly to intra-scientific sources of scientific change. Indeed the last-named book differs from traditional history of science less in the kind of explanation offered and more in the vast erudition and scholarly attention to detail displayed. Nevertheless, we should not think that Kuhn's practice is different from his preaching. The explanations Kuhn gives at the level of detail are consistent with, at a lower level of magnification, the framework of explanation being psycho-sociological. The explanation in terms of paradigms and exemplars is precisely that. The effect of exposure to exemplars is psychological. The training that exposes young scientists to the exemplars is sociological.

Is the symmetry postulate plausible? There are really two symmetry postulates – one for the pair true-false, and another for the pair rational-irrational. These need different treatment. Let us take true-false symmetry first. Newton-Smith points out that this version of the postulate is false, applied to low-level perceptual beliefs.[6] There is surely a difference in the explanation of perceptual belief between the individual who sees a chair in front of them and the other who does not but has a drug-induced hallucination of a chair. In the former case the chair plays a direct causal role and so it seems legitimate to answer the question *why did he believe the chair was there?* with *because it was there in front of him*, which will be a different sort of explanation from that employed in the hallucinatory case. But scientists are almost never in direct contact with the facts that make their theories true. What scientists are confronted with, what they reason from, and what they take as their point of origin is the evidence. Hence in the explanation of why scientists prefer one hypothesis to another it will usually be sufficient to refer to the evidence, not to the truth of the theory. That is not to say that the underlying facts are causally inert, for they may well be what cause the evidence to exist. But going to this level is usually irrelevant to the explanation of theory choice; if you ask scientists why they made this choice, then the answer will not be "because the theory is true", but rather "because the evidence is better explained by this hypothesis than the other, for these and for these reasons". So the symmetry principle is right as regards truth-falsity. But that is no reason to

215

adopt any sort of relativism about truth. On the contrary, it is because truth is non-relative that it can be hidden from us; and it is because the truth of scientific theories is hidden (to begin with at least), that this truth is irrelevant to *psychological* explanations of theory choice.

What then of the symmetry principle for rationality and irrationality? This seems an altogether different case. While the definition of rationality is even more troublesome than the definition of truth, whatever it is it should at least mark a difference that we all make quite naturally. Imagine someone believes with great confidence that he will win a lottery. We discover that he has only one ticket, that he has no reason to believe that the lottery has been rigged in his favour, that he knows that millions of tickets have been sold. Indeed he has no more reason to believe that he will win than anyone else has that they will win. Yet he has a firm conviction that he will win, one he demonstrates by big spending on credit in advance of the draw. Such a person we naturally describe as irrational. Why? Because all the evidence he has points to a very low chance of winning, one that justifies only a very low degree of confidence. We are thus inclined to look for a cause of his belief that is unrelated to the evidence – all too frequently people have a higher degree of belief in their chances in a lottery than is justified by the evidence because of the transient pleasure the thought of being rich gives them, and in some cases the degree of confidence caused is very high. This is clearly related to cases of self-deception, where a belief is motivated by a desire, a desire that gives no reason to believe. A mother who refuses to believe in the delinquency of her son, despite evidence that convinces impartial observers, has her state of belief fixed by the desire that her offspring should not be vicious. But a desire that it should be so is no reason to believe that it is so; desiring that *p* is no evidence that *p* is true. Other cases of irrationality include paranoia and other phobias. In all these cases we look for causes of belief that are outside the evidence, facts that lend no justification to the belief.

Newton-Smith argues that it is thus appropriate to look for explanations of irrational belief that are different from explanations of rational belief. Cases of irrationality do exist in science, most especially when scientists fail to give up belief in a favourite theory despite contrary evidence. Blondlot's belief in N-rays is best explained not by his rational response to the evidence but by self-deception. Frank Sulloway reports evidence from psychological studies that shows that a scientist's background, even whether he or she is a first or later born child, has an influence on his or her attitudes towards a hypothesis; for example, among Victorian scientists those who were first born were less quick to accept Darwinism than those who were later born.[7] It is likely that there are small tendencies to irrationality in all of us, scientists being no exception. Kuhn considers an instrumental, practical conception of rationality.[8] The rational choice of theory is that which enhances puzzle-solving power. If a choice of theory is motivated by self-deception or is to be explained as the outcome of a power struggle then the choice made may well fail to solve a greater number and variety of puzzles than its competitors. It will be irrational by these lights.

It is an empirical question how extensive the effects of irrationality are in science. On the one hand, research such as Sulloway's suggests that unalloyed reason might not be the norm. Furthermore, Kuhn's lesson about the functioning of paradigms as exemplars is that we need at the very least to be careful about what we mean by "rationality". If we think that rationality is characterized by certain rules of rational thinking, then much of the productive thought in science is not rational – it is quasi-intuitive instead. At the same time, such thinking is not irrational either; it does not go against what reason tells us. On the other hand, we might see rationality as more pervasive. We could have a concept of rationality that ties it less to rules and more to the mechanisms that generate justified belief. And if there are tendencies to irrationality, there are tendencies to rationality too; and the latter usually predominate, especially when it matters that we get things right. Even in Sulloway's Darwinism case the first borns came round once the weight of evidence was strong enough. Furthermore, as I shall suggest in the next chapter, the social organization of science, as described by Kuhn, is well placed to minimize the effects of individual irrationality. Science as an institution may be more rational than its practitioners. It seems therefore that in the explanation of the history of science we may find we need to appeal to irrationality less often than in everyday life.

There is therefore no requirement for an absolute symmetry in treatment between what we think of as rational and irrational thought. Nonetheless, we do need an active methodological principle of charity that enjoins us not to be too quick to treat as irrational the thinking of historical scientists whose beliefs we find difficult to understand. Careful attention to their paradigms may make it clear what they really meant and why it made sense to think as they did. Still less should we dismiss a belief as irrational just because it is

217

false. Later I shall suggest that a very mild form of relativism is appropriate for the concept of justification, and correspondingly for rationality too. The fact of disagreement among scientists is perfectly consistent with their all being rational.

Relativism and constructivism

Kuhn's evolutionary account of progress and his neutralism about truth do not require us to regard him as a relativist of any kind. His views are nonetheless sceptical, since he thinks that although we can rationally prefer one choice of theory to another we cannot think our preference takes us towards the truth. And, as I shall be arguing later, scepticism and relativism are closely related. In the Postscript 1969 he goes beyond his neutralism with a rejection of the concept of truth. Although this again is not relativism, it is one step closer to it, which Kuhn seems almost to acknowledge, for immediately after denying that he is a relativist, he says that his view has no need for the notions of truth or verisimilitude, ". . . if the position be relativism, I cannot see that the relativist loses anything needed to account for the nature and development of the sciences".[9] It is not surprising then that later he accepts a usage of "true" that is relative to paradigm-dependent lexicons.[10]

In this section I want to examine an influential argument for a kind of relativism that takes its cue from Kuhn and also from Ludwig Wittgenstein. Although Kuhn mentions Wittgenstein just twice,[11] it is true that there are other affinities between Kuhn's writing and Wittgenstein's later philosophy. The latter is usually regarded as reacting against many of the doctrines of logical positivism, as is the former. In particular, their remarks on perception, both of which relate to gestalt psychology, have much in common.

Kuhn's direct reference to Wittgenstein is an appeal to the latter's notion of a family resemblance concept. According to Wittgenstein objects may fall under a concept not because those objects all possess all of a set of defining properties, but instead because each object bears a similarity to some other objects in the set, a similarity that is not shared by *every* pair of the objects.[12] Objects not similar in this respect will be similar in some other respect. What binds these objects together under the one concept is thus a network of similarities, analogous to the similarities in appearance between members of the same family – some may have the same deep-green eyes, but others may not; those who do not have green eyes may yet share with some of the green-eyed members a characteristic nose, or with others a certain jaw-line. The concept "game" is supposed to be like this: there is no set of properties possessed by all and only things that are games. But there are salient resemblances between some pairs of games (they are played by teams, like rugby and bridge) that are not shared with other games (e.g. patience), while the latter will have a property in common with some other games (a property such as being played with cards, shared with bridge). Wittgenstein's point was that not every concept can be regarded as having its meaning specified or defined by a set of non-circular necessary and sufficient conditions for its application. Kuhn's point is an extension of this: some scientific concepts do not have non-circular necessary and sufficient conditions for their correct use; and thus we cannot specify explicit rules for the application of a theory containing such concepts to a new set of circumstances.

Some commentators have linked this to a particular interpretation of Wittgenstein's remarks on ostensive definition and rule-following.[13] Say we wanted to introduce a new predicative concept "C", by demonstrating some exemplary cases of being C, plus the stipulation that cases like the exemplars are C also. According to this *finitist* argument, *any* future use of the term "C" can be understood as logically compatible with the initial instruction. What does "like" mean here? Depending on one's perspective *anything* can be like (or different from) *anything else*. For example, we might try to teach someone "yellow" by giving them a sample of yellow canvas, and then telling them that something is "C" if it is like the sample. Someone might pick out as "C" just the things we regard as yellow, but there is nothing in the original instruction that rules out their thinking that something orange is to be called "C", or even describing a red object as "C"; after all, red things are like yellow things if one contrasts both to green or black things. At a greater extreme a piece of newspaper might be like the sample of canvas – it is certainly more like the sample than either is to a motor car. We come across other cases that turn not just on *degrees* of similarity: pigs, swordfish, swans and mussels might be regarded as similar to one another but different from goats, tuna, ducks and locusts if the context is that of providing kosher food. That perspective provides a basis for similarity that might not be apparent to those who lack that perspective. In turn this may suggest that we cannot rule out *a priori* the possibility of perspectives that permit apparently quite strange understandings of "like this", for example where only yellow things were "C" on even days of the month but pieces of canvas were regarded as "C" on odd days of the month. As Wittgenstein says of the attempt to fix the correct use of the term with such a rule "This was our paradox: no course of action could be determined by a rule, because every course of action can be made out to accord with the rule".[14] Kuhn seems to concur: "It is a truism that anything is similar to, and also different from, anything else".[15] One way of recognizing this is to imagine a society that regularly used terms like "grue".[16] For some deeply ingrained cultural reasons they might have a term "grellow" which we would define as "yellow on odd days, canvas on even days". Then the original sample would be grellow, and pieces of canvas on an odd day would thus be like the sample, both being grellow. We may have intended to define "C" as being yellow, but the original instruction was consistent with understanding "C" as grellow. Note furthermore that the original instruction is an example of the *use* of "C". Additional examples of the use of "C" might rule out grellow – we are told that a piece of red canvas on an odd day is not C. But the sort of reasoning given suggests that there are still a lot of weird interpretations of the original instruction that are consistent with countless additional examples of usage. Time-related interpretations in particular mean that any new use by the learner of "C" is consistent both with the instruction and with any past usage of the term. There are no facts, then, in past usage that determine some novel use of "C" as correct or incorrect. Consequently these facts determine neither a meaning for "C" nor whether "*x* is C" is true or false for some specific entity *x*. And what goes for "C" goes for any term of the language, not just new ones.

Be that as it may, we tend to think that words have meaning and sentences such as "sulphur is yellow" have truth values, and we are inclined to think that we know, often, what these are. But if these tendencies and inclinations do not come from a grasp of meaning derived from past use, whence do they originate? The finitist answer is that what makes some later application correct (or incorrect) is *community agreement* with that application (or disagreement, accordingly). Our training and induction into the community will give us expectations about what the community will say. In the scientific case the application of a theory to a new problem will not be determined solely by the exemplar but will also be dependent on community agreement. Consequently the scientific facts are not simply out there waiting to be discovered but instead come into being partly as a result of consensus.

There are two questions we may ask about such a view (a) does it have any truth in it? (b) does it have anything to do with Kuhn? The philosophical issues raised by (a) are considerable, and because my answer to (b) is "no", it is not necessary to respond to (a) in great depth. Nonetheless some indication of the problems with finitism is worthwhile and will also help see why Kuhn would not have endorsed it.

Scepticism and relativism about meanings ramify. Let us consider first the case of some object, x, that has not yet come to the notice of the community. Hence no consensus has been reached, at this time, that the sentence "x is C" is true, for any predicate "C". In the finitist view this lack of consensus means that the sentence "x is C" is not true; for that matter the finitist also holds that the sentence "x is C" is not false either. Barry Barnes seems to think that this is just a reflection on our established use of the word "true:" "'True' and 'false' are terms which are interesting only as they are used by a community itself, as it develops and maintains its own accepted patterns of concept application".[17] But things cannot be left there. For a key fact of the use of the term "true" is that it links propositions and the world, via the disquotational schema:

$$\text{"p" is true if and only if } p.[18]$$

For example: "Snow is white" is true if and only if snow is white.

Hence also: "x is C" is true if and only if x is C.

The finitist denies that the sentence "x is C" is true, and so denies the left-hand of this equivalence. Hence the finitist must also deny the right-hand side. The finitist must deny that x actually is C. This goes for whatever x and C are, if x has not been subject to examination and community consensus. So the finitist must hold that in advance of our examining them there are no roses that are red (or white or any other colour). Corresponding remarks may be made about the term "false", using the equivalence:

$$\text{"x is C" is false if and only if } x \text{ is not C.}$$

In the case of the unexamined object, lack of consensus on both "x is C" and on "x is not C" requires the finitist to deny that x is C and also to deny that x is not C. It is difficult to see how this escapes contradiction, since to deny that p is to assert that it is not the case that p, and so in this case we have the finitist asserting both that x is not C and that it is not the case that x is not C.

Barnes regards finitism as a social view of knowledge. But this is misleading. Issues of knowledge have not come into the picture yet. We have been discussing not the conditions under which someone knows that "x is C" is true; rather we have been asking the question, under what conditions *is* "x is C" true? Because of the disquotational schema, issues of concept application become metaphysical issues of the way things are. One cannot divorce the question "is it correct to apply the concept C to the object x?" from the question "is x C?" Barnes' view, whether he intends it or not, is primarily a social view of metaphysics, not of knowledge – although there will be consequences for knowledge too. It is what I have called strongly constructivist, since what things have what properties is dependent on, or constructed out of, a consensus that such things have such properties.[19]

Strong constructivism reverses the natural direction of fit. We would normally think, in a case of knowledge, that we reach a consensus that things are thus and so because they actually are thus and so; but on the constructivist reading, things actually are thus and so because we have reached a consensus that they are. This leads us to the second case, where a consensus has been reached that x is C. We may first ask what it is for the community to reach this consensus. If x's being C is a matter of the consensus existing, why should individual members judge that x is C? Indeed, immediately before the consensus they ought, if finitists, deny that x is C. So, in a community of finitists at least, it looks inexplicable that consensus should be reached on x's being C. Correspondingly, in communities where consensus is reached, that must be thanks to non-finitist thinking on the parts of the individuals; it must be natural for them to think that there is a fact whether or not x is C independently of consensus.

Apart from being metaphysically constructivist, Barnes' finitism is also metaphysically relativist. Since past use does not determine today's consensus, it is logically possible that the consensus will be that x is C and also logically possible that the consensus will be that x is not C. If the former possibility occurs then x *is* C, and if the latter x *is not* C. Hence which properties a thing has depends on which way the community swings. Now imagine that there are two communities, A and B, that swing different ways on this question. It follows that x is correctly called "C" and that x is not correctly called "C". Thanks to the disquotational schema this leads directly to a contradiction, that x is C and x is not C, unless somehow we regard such facts as being relative to some community.

Even so, a contradiction can be generated *within* a community. Individuals can change their minds. Therefore so can communities.

Hence the community on Monday may reach the consensus that roses are red. So, on Monday, roses are all red. Imagine that on Friday the community decides that it has erred, and so reaches the consensus that roses are not red, which means that on Friday roses are not in fact red. The only way to avoid the contradiction is to suppose that some roses have changed colour in the meantime. This is absurd enough, but consider had the community expressed its change of mind by reaching the consensus "We were in error on Monday in thinking that all roses are red." The finitist account requires then that it be the case that on Monday the community had erred, from which it follows that, on Monday, roses were not all red. But we have already seen that the community's consensus on Monday made it the case that on Monday roses were all red. Hence a contradiction is obtained – a contradiction not in the mind(s) of the community but in that of the finitist. It is the finitist's view of meaning and of truth that commits him to saying that it is true that roses were red and also true that roses were not red.

One way to avoid the conclusions of the last two paragraphs would be to claim that whenever there is community-level disagreement, the meanings of the relevant terms change, so the parties to the two sides of the dispute are no longer using the same language. This blocks use of the disquotational schema. This means that there is never genuine disagreement; one party cannot deny what the other denies as they are always talking past one another. While this is reminiscent of Kuhn's claims concerning incommensurability, in fact this finitist move goes far beyond it. For the finitist must think that this sort of incommensurability occurs *whenever* there is apparent disagreement, not just in cases of revolutionary theory change.

Kuhn certainly did not endorse the finitist's conclusions, nor did he support any of the premises that lead to them. Finitism is controversial even as an interpretation of Wittgenstein's remarks on rule-following.[20] His use of the family resemblance idea is generally regarded as quite separate, and it is only this that Kuhn explicitly draws upon. Kuhn's intention is to show that it is possible to possess a concept without knowing a set of non-circular necessary and sufficient conditions for its correct use, and the existence of family resemblance concepts would certainly help make the point. But the point does not require it to be the case that, in advance of consensus, *nothing* is a correct use, nor does it entail that consensus does fix a use as correct.

It is easy to conceive accounts of concept possession that satisfy Kuhn's point (and Wittgenstein's views on family resemblance) without approaching anything like finitism. As an example of such an account, to possess the concept C might be for someone to possess a certain disposition, so that to have the concept *red* would be to be disposed to use the term "red" of red things but not of things that are not red. There are numerous problems with such a view but the point of mentioning it is that it fits with Kuhn's view, since we can imagine that the disposition is acquired by an appropriate tuning of the neurophysiological system by exposure to samples of red and non-red things. We can see how it is possible to share concepts since our similar biological structures will allow us to be tuned to have the same dispositions if we share the same stimuli. The community has a role in ensuring that its members are all exposed to the same or very similar stimuli – which is Kuhn's point about shared exemplars.[21] Furthermore, this view is clearly not finitist, since dispositions extend beyond their actual manifestations; a sugar cube is soluble even if it is not actually dissolving; similarly I can be disposed to say of some rose that it is red even if I don't actually say that – if I haven't yet seen the rose. Hence a disposition could determine that it is correct to say that rose is red in advance of my, or anyone else's, coming to believe that it is.

Therefore, whatever the merits or demerits of the dispositional view just discussed, it is possible to endorse what Kuhn says without being a finitist. When we look at what he says in more detail it is quite clear that his opinions are not consistent with finitism. Kuhn understands Wittgenstein thus: "For Wittgenstein, in short, games, and chairs, and leaves are natural families, each constituted by a network of overlapping and crisscross resemblances. The existence of such a network sufficiently accounts for our success in identifying the corresponding object or activity. Only if the families we named overlapped and merged into one another – only, that is, if there were no *natural* families – would our success in identifying and naming provide evidence for a set of common characteristics corresponding to each of the class names we employ".[22] Kuhn is saying that it is because there are natural families that the possibility exists of introducing a concept without specifying necessary and sufficient conditions. Such concepts latch onto families and kinds that pre-exist those concepts. And so what makes it the case that something falls under a concept will not be consensus on that question but whether the object actually belongs to the family onto which the concept latches. Thus which concepts it is possible to acquire and which similarity relations one can perceive will depend on what kinds of things are actually there. Correspondingly certain salient similarities and differences will force themselves on us and be learned at some peril of doing otherwise. As Kuhn says, "An appropriately programmed perceptual mechanism has survival value. To say that the members of different groups may have different perceptions when confronted with the same stimuli is not to imply that they may have just any perceptions at all. In many environments a group that could not tell wolves from dogs could not endure".[23] Kuhn does *not* think that a creature would be a dog and not a wolf *just* because the community decided that it is, any more than a wolf could be Red Riding Hood's grandmother just by having a better disguise.

Kuhn's arguments against truth and verisimilitude

In the first edition of *The Structure of Scientific Revolutions* Kuhn's attitude towards truth is one of neutrality; he has no need of the concept, even in order to characterize and explain progress. In the Postscript to the second edition he introduces two arguments against the notion of truth implicit in the traditional view of progress as increasing verisimilitude. And yet later he relates his rejection of truth to incommensurability. Kuhn's remarks in the Postscript of 1969 are as follows:

> Often one hears that successive theories grow ever closer to, or approximate more and more closely to the truth. Apparently generalizations like that refer not to the puzzle-solutions and the concrete predictions derived from a theory but rather to its ontology, to the match, that is, between the entities with which the theory populated nature and what is "really there."
>
> Perhaps there is some other way of salvaging the notion of "truth" for application to whole theories, but this one will not do. There is, I think, no theory-independent way to reconstruct phrases like "really there"; the notion of a match between the ontology of a theory and its "real" counterpart in nature seems to me illusive in principle. Besides, as a historian I am impressed with the implausibility of the view. I do not doubt, for example, that Newton's mechanics improves on Aristotle's and that Einstein's improves on Newton's as instruments for puzzle-solving. But I can see in their succession no coherent direction of ontological development. On the contrary, in some important respects, though by not means in all, Einstein's general theory of

relativity is closer to Aristotle's than either of them is to Newton's.[24]

First, let us first look briefly at the historical argument. Note that it is in some tension with the first "metaphysical" argument. If the notion of "truth" is incoherent then how can history tell us that successive theories are failing to get closer to the truth? Furthermore, ought Kuhn to be confident of ontological similarity between Einstein and Aristotle given his warnings about the pitfalls of incommensurability? But for those who take truth seriously and who do not worry about incommensurability there is an important question concerning the historical evidence for convergent realism. Larry Laudan, who takes a problem solving approach to scientific progress that is similar to Kuhn's, argues forcefully that the evidence is against convergence.[25] It is significant that the best cases for the sceptical position come from the more metaphysical parts of physics. The "ontological" difference between Newtonian mechanics and general relativity is in the conception of space–time, and rather less in questions of the existence of gravity, mass, energy, momentum and so on. Yet for many philosophers the question of the nature of space-time is not even properly empirical but is partly conventional. Even if we grant Kuhn that point, he overstates his case when he says, "All past beliefs about nature have sooner or later turned out to be false".[26] There are many areas where the insights of earlier generations are not discarded but are instead refined or added to. In physics no one for many years has doubted Thomson's discovery of the electron, despite quite revolutionary developments in our beliefs about what electrons are. The same may be said for other particles discovered since. It is difficult to see that the basics of Dalton's atomic hypothesis could ever come under threat, even if the details might have changed. Nor need we fear falsification of subsequent discoveries concerning the composition of simple chemical compounds, for example that water is an oxide of hydrogen, or that benzene is a hydrocarbon with a ring structure. The quantum theory of the atom has changed our understanding of the chemical bond but not changed our belief in the existence of the atoms conjoined by it. The existence and structure of DNA is a fixture for all future science that descends from contemporary molecular biology. Harvey's discovery of the circulation of the blood is not likely to be revised, nor are our more general beliefs about the functions of the various major bodily organs. A realist would be foolish to maintain that all successions of theories always converge on the truth in all respects. But there is enough evidence for a strong case that it happens on a significantly frequent basis.[27]

Kuhn's first argument is somewhat obscure. What is the import of "There is . . . no theory-independent way to reconstruct phrases like 'really there'" Why does the phrase need "reconstructing"? Are we not clear as to what it means? Part of the point of the phrase is to point out that whatever we say or think in science, there is a basic sense in which what is *really* there is a separate matter. This use of "really" is no different from its use in sentences such as "many economists think that low interest rates cause inflation, but is that really the case?" If you can understand what that or any other such sentence means, then you understand what "really there" means. Just as there can be error in what people think, they can get it right, which is when what they say or think is true. Hoyningen-Huene understands Kuhn thus:

> The [. . .] argument is epistemological; it proceeds from the assumption that it's essentially meaningless to talk of what there really is, beyond (or outside) of all theory. If this insight is correct, it's impossible to see how talk of a "match" between theories and absolute or theory-free, purely object-sided reality could have any discernible meaning. How could the (qualitative) assertion of a match, or the (comparative) assertion of a better match, be assessed? The two pieces asserted to match each other more or less would have to be accessible independently of one another, where one of the pieces is absolute reality. But if we had access to absolute reality – and here we can only return to our initial premise – what interest would we have in theories about it?[28]

As to Kuhn's meaning I have no better suggestion than this, but if it is correct, then Kuhn's argument is extremely weak. First, why should knowing that there is a match between theory and reality require having independent access to each? Consider this analogy: I know that there is a match between my key and the levers of the lock, not because I have independent access to the levers of the lock, but rather because the key opens the lock. I shall return to this idea below pp. 237–40.

Secondly, the so-called "insight" is clearly false. As just pointed out, we have an intuitive notion of the possibility of error and of ignorance. And Kuhn must share this, since the only satisfactory explanation of the origin of anomalies is that the world is not exactly as our theories say it is. If error or ignorance can be shared by all of us, then there must be a way things are that is "beyond" theory. Kuhn is conflating metaphysical, semantic and epistemological questions here. Even if it were impossible to assess the assertion of a match, that would not make that assertion meaningless, unless one had some sort of verificationist view about meaning (another positivist trait). Where does the "insight" that it is meaningless to talk of what is there beyond theory come from? I suggest that this is associated with the true thought that for some bits of reality, if we can at all have specific thoughts about them or even knowledge of them, then that is possible only via theory. But from the fact that I can know about X only via theory it does not follow that no sense can be made of the thought that X might really be other than the theory says.

Far from the fact that I have access to (some part of) reality only via a theory being an obstacle to knowing the truth of the theory, it is precisely *because* the theory gives me access to reality that I can have knowledge of its truth. What a true theory asserts is just a description of some portion of reality. Hence knowledge of that bit of reality gives me knowledge of the truth of the theory. The point is perhaps most perspicuously grasped using the disquotational schema again:

(D)　"p" is true if and only if p.

of which one instance is:

(D1)　"Snow is white" is true if and only if snow is white.

and another is:

(E2)　"Protons are more massive than electrons" is true if and only if protons are more massive than electrons.

Let my theory be:

(T)　Protons are more massive than electrons.

Imagine that my experimentation and reasoning allows me to get to know (T) – I know that protons are more massive than electrons. Since (E2) is known to English speakers, I can easily know it too. From (T) and (E2) I infer,

(T*)　"Protons are more massive than electrons" is true.

Since (T*) in inferred by simple logic (Modus Ponens) from two propositions I know, I also know (T*). Knowledge of (T) is knowledge of reality, since protons and the like are components of reality. (T*) is definitionally equivalent to: (T) is true, and so knowledge of (T*) is knowledge of the truth of a theory. Hence knowledge of the truth of a theory can be had without some independent, theory-less access to reality with which to compare the theory. Kuhn seems to ignore the disquotational schema and the fact that the *point* of theory is to provide access to reality. It may be true that Kuhn is a sceptic who denies that we can have knowledge of theoretical reality (the world-in-itself). But that is no reason to misrepresent what would be involved in knowing the truth of theories were that possible.

We may be able to interpret Kuhn more sympathetically – perhaps more sympathetically than his brief comments deserve – as trading on a certain view of representation. Theories are representations; so all the intellectual things scientists do, such as wondering whether a theory is true, inferring a theory from evidence and so on, are just various kinds of thinking about representations. But our question: Is the theory true? depends on the way the represent*ed* is. Unless we have some independent access to reality it looks as if the best we can hope for is to deal with the representation only, and so leave the issue of truth untouched. On this view we conceive of theories as pictures and scientific thinking about theories as various kinds of investigations of the picture. If that were true, how could we ever hope to know whether the picture is a good representation of reality? The moral suggested is that we give up on any substantial notion of truth (and probably of representation too). Again this is to ignore the disquotational schema. The schema, which is a triviality for those that understand the language the theory is couched in, allows us to slip backward and forward between thinking about the theory (the represent*er*) and thinking about reality (the represent*ed*). There is no obstacle to moving from the question "is the theory 'protons are more massive than electrons' true?" to the question "are protons are more massive than electrons?" and back again. The questions may not be strictly the same in meaning, but a competent English speaker can see that whatever answer they require, they require the same one.

The mistake I am trying to diagnose arises if we regard the representation as having none of its connections with the world established in advance of the investigation into its fit with reality. It is as if we came across a theory written in a foreign language, to which, therefore, the disquotational schema is inapplicable. We could say: this is a theory; we don't know what it says; nonetheless, is it true? To get to know the answer we will typically have first to understand the language the theory is written in. And to understand the language is ipso facto to start making connections with the world. Thereafter, entertaining the various propositions asserted in the theory will be thought directly about the world.[29] This is the significance of reference. It is important not to misunderstand reference. Although referring to objects is often a matter of using names for them, we should not think of the act of naming as always involving taking an object in hand and baptizing or labelling it. If that were so, Kuhn would be right that one could not represent anything to which one could not have independent access. If one is an empiricist one does think of the root of meaning in this way, as forging a connection between words and independently accessible, i.e. observable, entities and properties. Hence the ability to refer to the unobservable is for the empiricist either a mystery or something that does not happen. It is symptomatic of Kuhn's residual empiricism that this is a worry for him too.

The key to avoiding such concerns is to realize that reference can be established without the namer or referer anything like literally pinning a label onto the referent. The term "neutrino" was introduced as the name of a certain kind of particle on the assumption that a kind exists that explains a certain phenomenon (a discrepancy in mass measured before and after beta decay). This naming was done long before any such particle was actually detected. Once reference was thereby established scientists could then refer to and have thoughts about neutrinos, including the creation of more complex theories about them. Note that for reference to be forged all that is required is that we have correctly identified a phenomenon that is explained by the existence of a unique kind particle. If that is so, then "neutrino" refers, which will remain the case even if we do not yet know this to be so. The conditions for reference are weaker than the conditions for knowledge. It is possible to refer to something even if we do not know it exists, and consequently we can be referring to things while ignorant of the fact that we are doing so; similarly we can think we are referring when indeed we are not, when the hypothesis that motivated the naming is false. Imagine Lavoisier hypothesizing about the existence of a gas that is part of the air and which combines with substances when they combust. This he calls "oxygen". We should grant that Lavoisier was indeed thinking and talking about oxygen (hence successful reference). But it may be that initially he also lacked sufficient evidence to know that his theory was true, and so did not know that oxygen existed. So Lavoisier was referring to oxygen ignorant of the fact that he did so, while Priestley thought he was referring to phlogiston even though there was nothing for him to refer to. The picture analogy for representation is inadequate to capture the way theories describe their subject matter. Hypotheses enable us to refer to unobservable entities. The names thus established may be used in the framing of further hypotheses, which can then support the introduction of new names for yet more deeply unobservable entities. I have admitted that we may not know these hypotheses to be true and so may not know that we have succeeded in referring. And certainly there is an epistemological concern that our ability to refer may outstrip our ability to know. But by the same token we cannot know that we have failed to refer unless we know that the supporting hypotheses are false. Scepticism about reference requires being unsceptical about theoretical knowledge.

I suspect that something analogous is to be found near the root of the finitist's view of meaning. The cases typically under discussion are those where the concept applies in virtue of unproblematically perceptible features of things. Colour concepts are typical cases. In many applications of a colour concept there is no reason to suppose that the individual fails to detect the relevant features. Let us call these "canonical" applications. Canonical applications show, in a direct fashion, the meaning of the term, or part thereof. A typical individual, Oliver, standing a couple of feet from a British postbox has no problems seeing it, its colour and so forth. Imagine Oliver says: "that colour is X" then I can tell something about what Oliver means by "X". If I follow enough instances of what Oliver says, then I may be able to predict what Oliver will say about things; I'll have a hypothesis about Oliver's meaning. If my predictions are falsified by some subsequent usage by Oliver, then I must revise my hypothesis. In canonical applications Oliver's utterances are always the arbiters of his meaning. We might now consider that meaning is a public phenomenon, and that what goes for Oliver goes for the community when it comes to the meaning of a word in the community's language. A community's consensus determines what their words mean.

This picture is too simplistic in important respects. First, it is an account not of what constitutes meaning but of how I might know what the meaning of a term is. Secondly, it holds only in canonical applications, that is in cases where the situation shows that perceptual or other cognitive error must be ruled out as an explanation of a particular utterance. But not all applications are canonical. If Oliver points to something in the distance, at dusk, illuminated by sodium street lights, and says "that is X" I am not forced to infer that "X" cannot mean "red" just because I happen know that the object pointed to is orange. I could reasonably infer that Oliver has made a mistake about the colour. Similarly, Oliver's statement that French postboxes are *X* need not falsify my hypothesis unless Oliver is reliably informed about the colour of French postboxes (they are yellow). Still less may I infer much about meaning from Oliver telling me that the colour of a certain star is *X*, if his pronouncement is the product of considerable theorizing, use of abstruse data and so on. Oliver could easily be wrong about the colour of the star. What goes for Oliver does indeed go for the community at large. The community's usage in canonical, optimal circumstances is strong evidence of their meaning, not constitutive of it. And in other circumstances, where there is more room for cognitive error, community consensus does not establish an application as correct – we could all be making a mistake. Furthermore, scientific circumstances are precisely those where the possibility of error is not negligible. Thirdly, not all concepts allow for canonical applications. It is perfectly possible to have a concept without there being the opportunity of reliably being able to apply it to any particular object. Theoretical concepts are typically like this, as is illustrated by the case of "neutrino".

Positivism and truth

Kuhn's argument against the correspondence conception of truth, for all its faults, has distinguished antecedents. Something like it may be found in Kant and also in James.[30] Revealingly, much closer precursors are to be found among the logical positivists, in particular Neurath and Carnap. Hempel describes the positivists' shift from a correspondence view to a coherence view, along with parallel changes in the nature of perceptual knowledge and observation that give us a striking anticipation of the structure of Kuhn's thinking on precisely these matters.[31] Hempel gives what he calls a "crude, but typical formulation" of Neurath's theses thus:

> Science is a system of statements which are of one kind. Each statement may be combined or compared with each other statement, e.g. in order to draw conclusions from the combined statements, or to see if they are compatible with each other or not. But statements are never compared with a "reality," with "facts." None of those who support a cleavage between statements and reality is able to give a precise account of how a comparison between statements and facts may possibly be accomplished, and how we may possibly ascertain the structure of facts. Therefore,

that cleavage is nothing but the result of a redoubling meta-physics, and all the problems connected with it, are mere pseudo-problems.

Neurath's worry about correspondence between propositions and facts, a central feature of Wittgenstein's *Tractatus Logico-Philosophicus*, is precisely Kuhn's, as represented by Hoyningen-Huene. "But how," asks Hempel, "is truth to be characterized from such a standpoint?" He continues:

> Obviously Neurath's ideas imply a coherence theory.
>
> Carnap developed, at first, a certain form of a suitable coherence theory, the basic idea of which may be elucidated by the following reflection: If it is possible to cut off the relation to "facts," from Wittgenstein's theory and to characterize a certain class of statements as true atomic statements, one might perhaps maintain Wittgenstein's important ideas concerning statements and their connexions without further depending upon the fatal confrontation of statements and facts, and upon all the embarrassing consequences connected with it.
>
> The desired class of propositions presented itself in the class of those statements which express the result of a pure immediate experience without any theoretical addition. They were called protocol-statements, and were originally thought to need no further proof.

This Hempel regards as the first step in abandoning Wittgenstein's correspondence theory. It replaces correspondence with "external reality" with a comparison to the basic components of experience. Then the second stage involves loosening Wittgenstein's post-*Tractatus* verificationist conception of meaning, so that universal statements, such as scientific hypotheses, can be regarded as having meaning even though they do not receive logically conclusive verification from singular statements. Furthermore, it is noted, even many statements that are singular in apparent form are hypothetical in logical form. Hempel says that the second step in departing from Wittgenstein "involves an essential loosening or softening of the concept of truth; . . . In science a statement is adopted as true if it is sufficiently supported by protocol statements."[32]

The third and final step in achieving the Carnap–Neurath conception of "truth" is to regard protocol statements not as absolutely reliable but as akin to the other statements of science as regards their revisability. It is thus that we reach Neurath's image of science as a boat that must be constantly rebuilt while at sea. There is no dry dock that allows rebuilding from the keel up.

I have commented already on the fallacious argument against the correspondence conception of truth. Even if truth is a matter of a match between statement and reality it does not follow that knowing a statement to be true requires direct comparison of both together. Why should not this match be known indirectly by inference of some sort? More significant is to see how this development of positivist thought leads to a position so similar to Kuhn's. As we saw in Chapter 4, Kuhn is sceptical about all except what scientists may observe with their eyes or instruments; statements of what they observe play in Kuhnian science just the role played by protocol statements in positivist science as Hempel describes it. Furthermore, the third step, rejecting the foundational reliability of protocol statements is parallel to Kuhn's claim of the theory-dependence of observation. While the grounds are slightly different, the resulting picture is just the same – while observation provides the basis for scientific belief, it is itself not immune from revision in the light of theoretical change. Once again it seems that Kuhn is not a radical anti-positivist. On the contrary, Kuhn is better seen as firmly belonging to the same tradition as positivists and empiricists such as Neurath, Carnap and Popper. Like them he rejects a set of claims specific to a certain kind of positivism associated with the followers of the early Wittgenstein, such as Moritz Schlick. Schlick's response to Carnap and Neurath is that their view leads to relativism about truth. It is perhaps worth recalling that one objection to the coherence view of truth is that many different, incompatible systems might be satisfactorily coherent. One response to this criticism might be to accept it, and so relativize "truth" to coherent systems. If we regard the shared beliefs of a normal science tradition as just such a coherent system then the relativized version of Carnap and Neurath's coherence conception gives us truth relativized to paradigms. A particularly striking parallel between these positivists and Kuhn comes in the response Hempel gives on their behalf to the question: What is it that tells us which system is the true system and which protocol statements are the true ones?

> The system of protocol statements which we call true, and to which we refer in every day life and science, may only be characterized by the historical fact, that it is the system which is actually adopted by mankind, and especially by the scientists of our culture circle; and the "true" statements in general may be

characterized as those which are sufficiently supported by that system of actually adopted protocol statements.

"How do we learn to produce "true" protocol statements?" asks Hempel.

> Obviously by being conditioned. Just as we accustom a child to spit out cherry stones by giving it a good example or by grasping its mouth, we condition it also to produce, under certain circumstances, definite spoken or written utterances, e.g. to say "I am hungry" or "This is a red ball."
>
> And we may say that young scientists are conditioned in the same way if they are taught in their university courses to produce under certain conditions, such utterances as "The pointer is now coinciding with scale-mark number 5" or "This word is Old-High-German" or "This historical document dates from the 17th Century."
>
> Perhaps the fact of the general and rather congruous conditioning of scientists may explain to a certain degree the fact of a unique system of science. [33]

Realism and the concept of truth

The positivist move away from the correspondence conception of truth is motivated by the same concerns as Kuhn's about "matching" with "reality" and the "facts". Two decades after the Postscript (1969) Kuhn writes:

> . . . what is fundamentally at stake is . . . the correspondence theory of truth, the notion that the goal, when evaluating scientific laws or theories, is to determine whether or not they correspond to an external, mind-independent world. It is that notion, whether in an absolute or probabilistic form, that I'm persuaded must vanish together with foundationalism. What replaces it will still require a strong conception of truth, but not, except in the most trivial sense, correspondence truth.
>
> . . . I earlier said that we must learn to get along without anything like a correspondence theory of truth. But something like a redundancy theory of truth is badly needed to replace it.[34]

For the positivists and for Kuhn the rejection of the correspondence theory is of a piece with their anti-realism. Indeed, it is often thought that because the correspondence conception and realism amount to the same thing, rejecting the one is tantamount to rejecting the other. This, I believe, is a mistake. Realism neither requires nor is required by a correspondence theory of truth. What is the correspondence conception of truth? It says that our statements have an underlying structure (their propositional structure) and that the world does as well. Certain rules, which we may understand as rules of meaning, correlate elements of the propositional structure and elements of the world. If these rules map the propositional structure onto the world structure, then the statement is true. If the two structures do not fit, then the statement is false. We can now see why this is independent of realism. On the one hand, the correspondence conception does not entail realism, since an idealist can interpret "the facts" and "the world" as consisting of mind-dependent entities such as ideas or sense-impressions. The idealist can then say that a statement is true precisely when its structure corresponds to the structure of one's ideas. According to some interpretations, Wittgenstein in the *Tractatus* took precisely this view. On the other hand, realism does not entail the correspondence conception, since realism is consistent with the minimalist account of truth. Minimalists note that the equivalence schema:

(E)   it is true that p if and only if p

is uncontroversially true for all propositions p.[35] The basic claim of minimalism is that an appreciation of this fact is all or almost all one needs for an understanding of the concept of truth. (One version of minimalism, is the redundancy theory to which Kuhn refers. It claims that the schema (E) shows the concept of truth to be redundant – it can be straightforwardly eliminated in all contexts. This seems to be over-optimistic. But even if use of the concept is unavoidable in certain contexts it may still be the case that no substantive theory is required to explain it and that the simple equivalence principle, given by schema (E), is sufficient.[36]) Minimalism is consistent with realism since realists can give a realist interpretation of the content of p. Thus one instance of (E) is:

(E1)   it is true that protons have a greater mass than electrons if and only if protons have a greater mass than electrons.

According to the realist the content of the right-hand side, "protons have a greater mass than electrons", includes the claim that a certain unobservable entity has more mass than another unobservable entity.

Furthermore there is nothing in the minimalist conception that rules out knowing that protons have greater mass than electrons. So minimalism is consistent with realism.

Given that one can be a realist without endorsing any correspondence theory, what is the connection between the rejection of the one and the rejection of the other? To see the connection we need to remind ourselves of the detail of the reasons given by Kuhn and the positivists for rejecting the correspondence conception. Talk of "matching" and "comparing" theories and reality gives the impression that Kuhn and the positivists are objecting to the correspondentist idea of structural isomorphism. But in fact they have nothing to say about that idea in itself. Their complaint is that we could not *know* of the existence of such a match, because we could not have theory-less access to reality. The confusions in that complaint have already been analysed. But for present purposes we should note that the complaint is an epistemological one (as Hoyningen-Huene notes), concerning knowledge of reality, not a metaphysical one concerning structural isomorphism. From this I draw two conclusions. First, for Kuhn and the positivists alike their sceptical anti-realism is not a *consequence* of their rejection of the correspondence theory; instead the relation is the reverse – they reject the correspondence theory because of an antecedent anti-realism expressed as a concern about our access to reality. Secondly, the rejection of the correspondence theory is really a rejection of something different. It is the rejection of any view that holds both (a) that the notion of truth is to be explained in terms of the way things are rather than the reverse, and (b) that the way things are need not be knowable directly (e.g. by perception). The correspondence conception is committed to (a), but so also is the minimalist conception, since in (E) the right-hand side explains the left-hand side. Realism gives a commitment at least to (b), and, I think, to (a) as well.

Verisimilitude and incommensurability

Kuhn has another objection to verisimilitude whose source is the incommensurability of theories. He says that incommensurability, the no-overlap principle in particular, prevents us from expressing the propositions of Aristotelian physics and those of Newtonian physics in the same language. "It follows that no shared metric is available to compare our assertions about force and motion with

Aristotle's and thus to provide a basis for a claim that ours (or, for that matter, his) are closer to the truth."[37] We have discussed the problem of incommensurability at length. One conclusion is that there is no general reason to suppose that there cannot have been continuity of reference between Aristotelian and Newtonian physics; and so it may be possible to identify pairs of propositions such that one is true when the other is false and also such that one is a consequence of Newtonian physics and the other a consequence of Aristotelian physics. Here I want to pose a different question for Kuhn. Why should we need to identify such pairs of propositions for an assessment of verisimilitude to be possible? Kuhn is encouraged both by Popper's account of verisimilitude and also by empiricist notions of empirical confirmation to think that if one were to establish relative nearness to the truth one would have to compare theories in a point by point manner.[38] We saw that the Popper's account involved one theory getting things right where the other one got them wrong. We saw also that confirmation for empiricists is a matter of a hypothesis possessing verified consequences. It is natural then to think of relative nearness to the truth in this way: if theory A has a consequence p and theory B has consequence not-p then, as far as this pair of propositions is concerned, A is nearer to the truth if p is true, while B is nearer if p is false. Overall relative verisimilitude depends on how matters stack up for all such propositions.

Again this is a case of Kuhn's conceptions being informed and limited by his empiricist inheritance. Rather than seeking to comprehend truth-likeness and confirmation in non-empiricist terms, Kuhn accepts the empiricist understanding of them, and so rejects possibility of knowledge of nearness to the truth altogether. That is a mistake. Theories are compared to one another. But they are not always compared in a point by point fashion. For example, there are various hypotheses that explain the mass extinctions of dinosaurs: evolutionary unfittedness, volcanic eruptions and meteor impact among others. As regards consequences we might observe that these hypotheses generate relatively few that directly conflict with one another. What sort of considerations are relevant then? The evolutionary hypothesis has the advantage that we know that the process of extinction by evolution occurs; but it is weak nonetheless because it seems to leave unexplained the catastrophic nature of the mass extinctions: very many species died in an evolutionarily short space of time. The meteor hypothesis is regarded by many as the strongest because it explains anomalously high deposits of iridium in rock strata associated with the extinctions. Iridium is normally rare on Earth but is more highly concentrated in meteors. Neither the volcano hypothesis nor the evolutionary explanation says that there should *not* be unusual iridium deposits. Some supporters of the volcano view say that they too can account for the deposits. What this brief example suggests is that hypotheses are compared with respect to their ability to explain various salient pieces of evidence. Our preference for one hypothesis over another reflects a belief that it is overall a better explanation of the evidence than its competitors. The exact details of the structure of this kind of scientific inference – Inference to the Best Explanation ("IBE") – are subject to some debate.[39] Nonetheless, details notwithstanding, IBE is an account of theory comparison and preference that does not require point by point comparison of the competitors. It might be thought that what is being decided here is the correct solution to a puzzle, not the comparison of theories from different paradigms. Yet to many the meteor hypothesis is revolutionary (it is certainly revisionary). And even if this is just normal science there is no reason to suppose that IBE is out of place in extraordinary science.[40] Indeed, from the perspective of revolutionary change, IBE has several things going for it. It treats theories not piecemeal but as packages, since explanations are typically integrated sets of propositions not single ones. Hence we see whether one theory, taken as an integrated explanation, is a better explanation than another, also taken as a single package. To some extent this respects the holism Kuhn identifies. Furthermore, seeing whether a theory as a whole is a good explanation requires us to appreciate a theory in its own terms. A theory may introduce its own lexicon and taxonomy and we may use this in seeing whether the theory is any good, and we may shift to the terminology of the competitor theory when examining it. We do not need to translate between the languages of "phlogiston" and of "oxygen" to see that in its own terms phlogiston theory is a poorer (or stronger) explanation of the evidence than oxygen theory is in its terms – we do not have to have one all-encompassing, uniform language suitable for expressing all theories.

Reflecting on IBE might help us in exorcising some of the misconceptions about truth shared by Kuhn and the positivists. If we start with pure "observation" statements, statements about the objects of direct perception (which for some empiricists may be inner states), then the (early) positivist view of the assessment of their truth is straightforward. The content of the statement is compared with the independently given data of perceptual experience. A match between the statement and what is experienced gives truth, otherwise the statement is false. This is then elevated into an assumption as to what the correspondence conception of truth requires for all assertions, whether observation statements or not. It is easy to see where the thought originates, that knowledge that a theory is true would require theory-less access to reality. To counter this line of thinking we should consider that the manner of knowledge acquisition exemplified by the observational case may be a peculiarity of perception and not anything to do with the concept of truth (conceived as correspondence or not). Inference may give us knowledge of the truth of propositions also, but since it is *inference* that knowledge is not knowledge by direct confrontation of statements and an independently accessed reality.

Epistemological incommensurability

Although semantic incommensurability came to dominate Kuhn's thinking, it is not the only kind of incommensurability to be found in *The Structure of Scientific Revolutions*. There he also sketches an epistemological incommensurability that implies a version of epistemological relativism, the view that the epistemic evaluation of a proposition is at best relative to a variable set of standards such as a paradigm.[41] In this section I shall outline the sort of case that can be made for such a view and in subsequent sections I shall examine and challenge the assumptions underlying it.

Kuhn says very little about traditional concerns in epistemology, which may initially seem surprising for the author of a book on theory development and change. This is explained by his neutralism and rejection of the concept of truth. Standard epistemology, as remarked, takes knowledge to be knowledge of truths, where "truth" is understood as truth that is not relative but absolute. One cannot know a proposition that is false, although one can mistakenly believe it. If a person knows that p then it is true that p – knowledge is *factive*. If truth makes no difference to our explanations in the history of science, then knowledge will not either. Insofar as Kuhn extends his rejection of empiricism to an attack on the notion of truth he threatens to remove the possibility of *any* standard kind epistemology, not just empiricist epistemology. Furthermore, the sort of truth-relativism Kuhn adopts in the later papers makes it difficult to be even an epistemological relativist (let alone an absolutist). For epistemic relativism implies that one and the same thesis may be evaluated by different sets of standards (which therefore may yield different evaluations). But Kuhn's later position denies that a thesis framed in the lexicon of one paradigm is even a candidate for evaluation by a paradigm that employs a different lexicon. So Kuhn's later taxonomic incommensurability simply undercuts the issue of epistemological relativism or absolutism.

But this was not always Kuhn's position and there is strong evidence of epistemological relativism in *The Structure of Scientific Revolutions*. There Kuhn wrote:

> when paradigms change, there are usually significant shifts in the criteria determining the legitimacy of both of problems and of proposed solutions . . . to the extent that two scientific schools disagree about what is a problem and what a solution, they will inevitably talk past one another when debating the relative merits of their respective paradigms. In the partially circular arguments that regularly result, each paradigm will be shown to satisfy more or less the criteria it dictates for itself and to fall short of those dictated by its opponent . . . since no paradigm ever solves all the problems it defines and since no two paradigms leave all the same problems unsolved, paradigm debates always involve the question: which problems is it more significant to have solved?[42]

The basic idea is clear. Paradigms are governed by their exemplars; exemplars are what define the puzzles for a paradigm and the standards of solution. When a paradigm changes its exemplars change and so do the relevant problems and the criteria for solving them. What counts as a good problem in one paradigm need not count as good problem in another: a puzzle that seemed worthwhile by virtue of similarity to the earlier exemplar may now be deemed irrelevant on account of its dissimilarity to the new exemplar. And where there are problems in common, what counts as a good solution may differ between the paradigms: a proposed solution may resemble the exemplary puzzle-solution of one of the paradigms but not that of the other. Kuhn acknowledges that there will be similarities in the standards of evaluation, in particular the values of accuracy, consistency, breadth of scope, simplicity and fruitfulness which will be shared by different paradigms. But even these paradigms may differ in the application of these values: they may disagree as to what counts as simplicity or fruitfulness, what was complex before may now seem simple and vice versa. And even if there is agreement as to the assessment of the individual values there may be no consensus on which values are the most important.

The implication seems clear: standards are at best relative; there are no absolute, external standards of evaluation. Correspondingly the epistemic outcome of evaluation – knowledge, justification, rational belief etc.– must at best be relative. Old Rationalists believed in the existence of a general, *a priori* justifiable scientific method. They disagreed about what exactly it is – whether inductive logic, hypothetico-deductivism, or falsificationism – but then it is the job of philosophical theorizing and debate to decide that question. Kuhn's account of scientific change may be seen as undermining this view. Unlike the scientific method of the Old Rationalists, the standards of normal science are not universal, let alone *a priori* justifiable. They can be shown by the growth of anomaly to be inadequate and will be replaced with something different when a revolution occurs. As a result of the revolution and a change in exemplar a new standard is set for the adequacy of hypotheses. We have a non-semantic version of incommensurability: incommensurability due to variance in standards.

Since proposed puzzle-solutions devised under a particular paradigm will aim to be similar to its exemplars, they are thus likely to have some properties in common with the exemplars, even if they are poor solutions. With these exemplars at hand we may be able to judge how closely a puzzle-solution gets to being like the exemplar. But a puzzle-solution from one paradigm may not even have the appropriate properties that would allow its evaluation in another paradigm. For example, Copernicus sought to respond to the criticism that if the Earth rotated, the forces of rotation would destroy it, with the following argument:

> But if one holds that the Earth moves, he will also say that this motion is natural not violent. Things which happen according to nature produce the opposite effect to those due to force. Things subjected to violence or force will disintegrate and cannot subsist for long. But whatever happens by nature is done appropriately and preserves things in their best condition. Idle, therefore, is Ptolemy's fear that the Earth and everything on it would be disintegrated by rotation which is an act of nature, entirely different from an artificial act or anything contrived by human ingenuity.[43]

But this, to modern ears doesn't even sound like an argument. It seems not so much bad physics as not anything recognizable to us as physics. And this is because Copernicus had Aristotle, who similarly made much use of a distinction between natural and violent motions, as his model.

A musical analogy may be instructive. We are quite happy judging Schubert to be a great composer and his contemporary Löwe as being a bit better than mediocre. This is because we are familiar with the standards that apply to such music, we know their models and understand what they were aiming at. But such knowledge, however detailed, would not enable someone unfamiliar with Indian classical music to make a judgment of the quality of the music of the composer Tyagaraja nor even whether it is better than that of Syama Sastri (both being contemporaries of Löwe and Schubert). And even if we are connoisseurs of Indian music we may feel that to be asked to rate all four composers on a single scale is to ask something pointless, impossible or even senseless. And a major part of the reason for this is that while Schubert and Löwe were aiming to achieve more or less similar goals, Tyagaraja was clearly aiming at something quite different, the different goals being set by different traditions. Mozart, for example, had shown how remarkable subtlety in harmonic development was available within the rules of classical harmony, effects which his successors sought to emulate. Given Mozart we know what to look for in those successors. We find signs of the attempts and see that Schubert manifestly equalled and perhaps surpassed Mozart in this respect; we see similar signs in Löwe of his aiming at the same thing, and are thus able to judge that he failed to get close. But since Tyagaraja was not aiming to match Mozart nor to operate within the rules of classical western harmony, there is nothing we could latch onto in order to start making a judgment. In the scientific case we may be happy about judging the quality of a puzzle-solution that aims at similarity to a familiar exemplar. But if it does not even attempt to be similar to our exemplars it will not have the features we look for in starting to make a assessment. For these reasons if we are operating within a particular paradigm we can judge the quality of a puzzle-solution generated by that paradigm, but if we see a solution generated by another paradigm we will find it hard even to start assessing it; it may not even look like an attempted puzzle-solution. Thus, in some cases, the problem is not so much that we will judge a puzzle-solution to be of low quality that another paradigm judges to be of high quality, but rather that we do not know how to judge it at all.

In practice puzzle-solutions from successive paradigms will not differ quite as radically as classical European and classical Indian music. There is some continuity through a revolution and so there are likely to be many features in common between puzzle-solutions from either side of it. Even then there may be subtle differences. One of the features of a disciplinary matrix is the shared "values" of the normal-scientific community. I explained that values, like other aspects of the disciplinary matrix, are dimensions of potential resemblance between an exemplar and a puzzle-solution. So the value of simplicity is a feature that can be held in common by an exemplar and a satisfactory hypothesis. An exemplar in one paradigm may show some such feature, the exemplar of another may not – that feature will be a virtue of potential puzzle-solution operating within the former paradigm but not within the latter. As a matter of fact Kuhn thinks that certain values are ubiquitous in science – accuracy, simplicity, fruitfulness, consistency and breadth of scope. But even then there is room for difference over what counts as simplicity or fruitfulness and for difference over how these values should be ranked.[44]

The possibility of incommensurability of this kind need not lead in every case to non-comparability. There may be sufficient similarity between successive paradigms and sufficient dissimilarity between a pair of puzzle-solutions that both paradigms lead to the same assessment. But it must in some cases lead to non-comparability, and quite frequently in important cases. This is because the very point of the exemplar's being a source of standards and values is that this is how puzzle-solutions are evaluated for their goodness. It is on the basis of such evaluation that puzzle-solutions are adopted or rejected. Consequently when there is a difference between exemplars there must be a difference in evaluation for some possible puzzle-solutions. Furthermore, since a revolution occurs when an earlier paradigm that fails to solve problems is replaced by one that succeeds in so doing, it follows that the new puzzle-solutions that are positively evaluated by the new paradigm would have been evaluated as poor by the old paradigm. If the old paradigm would have given them a good evaluation then no revolution was necessary. Because progress through revolutions is a matter of increasing problem-solving capacity, not increasing proximity to the truth, we cannot in Kuhn's view say that evaluation by the later paradigm is more reliable or accurate than by the former. The musical analogy is apt here. There is a sense in which classical music has made progress. The expressive resources available to modern composers are wider than those that were available to their medieval predecessors. But that fact does not translate into the claim that modern pieces of music provide better or higher standards of comparison for the evaluation of musical quality than the masterpieces of the *ars antiqua*. It is only when there is progress towards some normative goal, such as truth, that later instances can be regarded as providers of better standards than earlier ones.

Externalism and internalism in epistemology

Barnes' finitist conception of meaning implies a relativist notion of truth. Whether a statement is true depends on a consensus among the users of the language in question. Let us then distinguish between R-truth (relative truth) and A-truth (absolute truth). Let us call the standard conception of knowledge "A-knowledge". A-knowledge entails A-truth. The standard view of knowledge is typically a part of empiricism, but it is also consistent with many non-empiricist epistemologies. A non-standard epistemology denies that "knowledge" entails A-truth. An epistemology may be non-standard because it rejects A-truth; or it may accept A-truth but employ a notion of R-knowledge where "S R-knows that *p*" does not entail that *p* is A-true.

Kuhn's view of perception is non-standard. As we saw in Chapter 4 Kuhn regards "S sees that *p*" as non-factive. Since "S sees that *p*" entails "S knows that *p*" we may infer that Kuhn has a non-factive notion of knowledge.[45] The epistemological version of incommensurability discussed in the last section appears to rule out A-knowledge. How can A-knowledge be generated by hypothesis assessment based on comparison to a contingent, changeable exemplar, one that may be taken in time to be refuted by weight of accumulated anomaly? The challenge is that A-knowledge cannot be the product of a science for which Kuhn's historical description and theory are true. Hence we should either be sceptics about scientific knowledge or should make do with the weaker notion of R-knowledge. Similar remarks may be made with regard to other epistemic notions such as justification and rationality.

To see why I do not think that there is a serious problem of incommensurability due to variance in paradigms and standards, it is necessary to explain certain recent developments in (standard) epistemology. The view I wish to expound comes under the headings *naturalized epistemology* and *externalist epistemology*. Not every view which comes under these headings will agree with mine in all details.[46] But, I suggest, as regards a response to incommensurability of standards, all these views can agree in essentials. These views contrast with epistemological *internalism*. Internalism has been dominant in epistemology at least since Descartes, especially where epistemological reflection has lead to scepticism.[47] Take Descartes' epistemological device of the evil demon. The evil demon constantly deceives us, giving us perceptual experiences which are plausible but in fact false. The sceptical argument proceeds thus. Our perceptual processes might be reliable and veridical but they might also be controlled by the evil demon. There is, says the sceptic, no way of telling which. Since we don't know whether our perceptual processes are reliable or deceptive, the perceptual beliefs we have as a result of using those processes cannot themselves count as knowledge. According to this argument it is necessary in order for S to get knowledge from her perceptual processes that she also knows that her perceptual processes are reliable. Thus the key issue in determining whether S is in some epistemic state is *always* a matter of S's other epistemic states.[48]

Put another way, on this internalist view, S's epistemic states are always known to S – S's states of knowing and justification may always be determined *internally* by S. Let it be the case that S does know that p. The internalist reasoning of the previous paragraph says that in this case S must know that the process by which S formed the belief that p was, on this occasion, sufficiently reliable to produce knowledge. So S knows that she knows that p. In short, if S knows that p then S knows that S knows that p. If we symbolize "S knows that x" by "Kx" then we may symbolize the principle stated in the last sentence so: $Kp \rightarrow KKp$. The principle is thus known as the "K-K principle".

According to the view of epistemology I am sketching here, the internalist is committed to the K-K principle, and a similar discussion couched in terms of justified belief will lead to demonstrating the internalist's commitment to a J-J principle: if S is justified in believing that p, then S is justified in believing that S is justified in believing that p. There are good reasons for rejecting the K-K and J-J principles, and so the internalist epistemology of which they are a part. Firstly, there are strong arguments that show that these are simply false.[49] Secondly, our everyday usage of epistemic terms does not conform to the principles. We acquire beliefs as the result of the operation of certain processes, procedures, methods and so on: perceptual processes (seeing, hearing etc.), measuring procedures, experimental methods, inductive reasoning of various sorts, observation using instruments, mathematical and logical deduction and the like. In some cases, such as mathematical and logical deduction, we may be able to establish through pure reason that they will yield some knowledge. But in the case of perceptual processes, experimental methods and inductive reasoning the reliability of these cannot be known *a priori*. Outside philosophy at least, we do not regard knowledge of our own processes as a precondition of using them to give us knowledge or justified belief. There are some individuals who have never reflected on the reliability of their eyesight and have never had an eye test. But we would not want to say that therefore they cannot by looking know what is in their immediate environment. Small children and animals are obvious cases of such individuals. I should add that many adults are probably in this position too, although it will be responded that most adults have a fair idea of the quality of their eyesight simply from the everyday fact of finding their visual judgments confirmed or disconfirmed by other or later experiences (e.g. hearing and touching what they see, or by closer visual inspection). Nonetheless we may imagine the case of someone with excellent eyesight suffering from amnesia. Such a person would not be able to tell you whether his eyesight has a good track record. But that does not prevent his being able to know by looking that he has awoken in a room with white walls and in a bed with red blankets.

We can say similar things about other sources of knowledge, for instance testimony – knowledge and belief formed on the basis of reports from other people, books, newspapers etc. You have a lot of general as well as specialist knowledge gained this way. Can you remember where you learned it all? Probably not. And in the cases where you can remember, say, reading it in a newspaper, did you, at that time, know of the reliability of the reporter who wrote the article? And if you did, where did this knowledge come from? Did you know of the reliability of its source? If people get to know things because you have told them so, do they check on your reliability? And on the reliability of your sources? If knowledge from testimony required prior knowledge of the reliability of the source of the infor- mation, and knowledge of the reliability of its sources and so on, we would have very little general knowledge at all. But we do not think this is the case. Perhaps we have less than we think – but not that much less. The naturalized epistemologist concludes that it is neither part of our ordinary concept of knowledge, nor of our concept of justification, that in order to have knowledge or justified belief we must have prior knowledge or justified belief in the reliability of our processes, sources, procedures or methods.

It is possible to respond to these remarks on everyday usage by saying that the latter is mistaken: we should be sceptical as regards A-knowledge and A-justification. The epistemological externalist denies that we are systematically mistaken in our attributions of knowledge; it is just as a feature of our epistemic concepts that they do not respect the K-K and J-J principles. This is the first and most important conclusion of externalist epistemology. We can know that *p* even though we do not know the reliability of the process by which we acquired the belief that *p*; similarly we can have a justified belief that *p*, even though we lack a justified belief in the reliability of the process leading to the belief.

It is thus the *actual* reliability of the belief-forming process that is important for knowledge, not that one knows the process to be reliable. Take the amnesiac again. He may not be able to tell us whether his eyesight is good or not. But it is important whether it actually is good or not. If as a matter of fact his concussion has impaired his colour vision, say by making him colour blind, then he won't be able to know what the colours of things about him are – he may judge that the blankets are red, and he will be right, but he won't know that they are red since his colour blindness means that he cannot tell red blankets from green ones.

The significant issue, therefore, in fixing whether the subject S, who believes that *p*, thereby knows that *p*, is the *actual* reliability of the process which formed that belief, *not* S's knowledge (or justified belief) concerning that reliability. This is why this approach to epistemology is called *externalist*. They key issues in fixing S's epistemic state when S believes that *p* are (typically, though not always exclusively) *external* to S's other epistemic states.

This is not to say that we cannot have knowledge that we know or knowledge of the reliability of our belief-forming processes. Often we are able to say something about the reliability of our belief-forming processes. So, for instance, someone may be able to tell me that she has good eyesight, because she has never had trouble seeing things, has never been led astray by her eyes, and has recently had her eyes tested by an optician who says that they are in excellent condition. What is important, however, is that this second-order knowledge is not of a special kind and is not acquired in a special way. Implicit in internalist, Cartesian epistemology is the thought that the questions "What do we know?", "How do we know?" are special, philosophical questions to be answered by philosophical methods quite unlike the methods by which the first-order knowledge is generated. In the naturalized view these questions are just like any other questions of contingent fact and should be investigated in the same kind of way. This is what is meant by saying that epistemology is naturalized; some parts of epistemology are no longer regarded as parts of philosophy but instead as parts of ordinary enquiry and of the natural sciences in particular.

In many way this should not be surprising. Knowledge is after all a natural phenomenon. It is not only humans that know things; other creatures with perceptual systems know things too. And this is because evolution has equipped us with capacities that are responsive to our environments, including the capacities for having experiences and of forming beliefs. Clearly there is a selective advantage in having the capacity to form true beliefs as opposed to a capacity that gives false beliefs.[50] There are evolutionary explanations of our capacities for knowledge as well as our capacities for walking upright and manipulating tools. Evolutionary biology is the appropriate science for explaining why in general we have reliable belief-forming processes. If we ask which specific belief-forming processes are reliable and why, we need to consult different sciences. Thus cognitive science and neurophysiology tell us about our perceptual capacities, how they work and where the limits on their reliability lie. Our belief-forming processes go beyond the merely perceptual. We often use instruments, especially in science. In these cases we typically do not investigate their reliability after their invention but before. Instruments are designed in order to be reliable, to give knowledge. These designs will depend on theories, and so our knowledge of the reliability of the instruments will again be *a posteriori*, theoretical knowledge. Instruments are not the only extension beyond unaided perception. Testimony is another sort of case. Of course testimony all too often comes from someone who is unreliable and so what we are told we do not know to be the case. Sorting the reliable from unreliable sources (friends, teachers, newspapers, books, broadcasters etc.) is a task we begin to learn from a young age – it is a little piece of naturalized epistemology. Sometimes there are social institutions that ensure high levels of reliability and it may be the task of sociology to identify them and explain why they are reliable (we shall return to this).

Not only may the sciences and other forms of investigation tells us who knows what and how, they may also help us to improve the methods and processes we use so that we may get to know more. This is already suggested by the use of theory in the design of instruments. A different kind of case is the introduction of double-blind testing in medical science. Cognitive psychology can also tell us which ways of forming beliefs are unreliable: research suggests that interviews are often poor ways of getting to know whether a candidate is suitable for a job and well-informed firms have adjusted their recruitment policies accordingly. The eponymous litmus test in chemistry, Koch's postulates in microbiology, and the hundreds of tests that are carried out mechanically in medical diagnostics are further examples of belief-forming procedures that are the products of science. The view, held by Kuhn and Feyerabend, that there is no single, epistemologically privileged scientific method is fully endorsed by naturalized epistemology. Instead there is a motley of rules of thumb, inference procedures, methods and processes that play a part in forming belief. Furthermore, it is one of the roles of science to add to them.[51]

Since such investigations of knowledge are natural, typically *a posteriori* investigations, they do not have any special authority. Like inquiry in general these investigations can go wrong. They themselves can employ unreliable methods, and those who undertake them will fail to know who knows what and so on. In general it is easier for S to know that *p* than for R to know that S knows that *p*, since for the former S need only *use* a reliable method, while for the latter R must *know* that S's method is reliable. In particular S more easily knows that *p* than knows that she knows that *p*. Imagine that S knows that *p* but does not know that she used a reliable method or process to acquire that belief and so does not know she knows that *p*. S is therefore in no position to persuade T rationally that S knows that *p*, unless T additionally knows something that S does not. In general, therefore, knowing that *p* need not be sufficient for persuading someone else that one knows that *p*. Imagine also that T doubts that *p* and furthermore doubts that S's methods are reliable. Since S does not know that her methods are reliable it follows from what we have just seen that S may not be able to get T rationally to believe that her methods are reliable. Since T is not rationally persuadable that the methods are reliable, T may not be rationally required to use them. Consequently T may not be in a position to get to know that *p*. Thus even though S knows that *p*, S may not be able to get T to know that *p*, however rational both S and T are. As an illustration, imagine that S has been inaccurately represented as a liar to T by people whom T has good reason to believe are reliable judges of character. S may know that *p*, but T may even be rationally required *not* to believe that *p*, whatever S says. Knowledge is not always transmissible.

Scepticism, relativism and internalism

In the next section I shall show how the machinery of externalist, naturalized epistemology undermines the claim that incommensurability due to variance of standards leads to scepticism or relativism about knowledge. In this section I shall further prepare the ground by providing some analysis of the relativist's position. I shall start by emphasizing that relativism is a version or off-shoot of scepticism. We have seen an indication of this already. Kuhn, as we have seen, thinks that A-truth is a concept without coherent use. This is a sceptical view. Nonetheless he does say that we may use the words "true" and "false" relative to the lexicons associated with different paradigms. The lexicons are not themselves true or false.

> A lexicon or lexical structure is the long-term product of tribal experience in the natural and social worlds, but its logical status, like that of word-meanings in general, is that of convention. Each lexicon makes possible a corresponding form of life within which the truth or falsity of propositions may be both claimed and rationally justified, but the justification of lexicons or of lexical change can only be pragmatic. With the Aristotelian lexicon in place it does make sense to speak of the truth or falsity of Aristotelian assertions in which terms like "force" or "void" play an essential role, but the truth-value arrived at have no bearing on the truth or falsity of apparently similar assertions made with the Newtonian lexicon.[52]

Given a lexicon within which a statement is a "candidate for true/false", what makes a statement rationally assertable are the normal rules of evidence.[53] This is a relativist view, in that while "true" cannot be used inter-theoretically it may be used intra-theoretically, without its use with regard to statements in one theory impinging on its use with regard to statements from in connection with another theory.[54]

Similarly the epistemic relativist argues that for various reasons A-knowledge is impossible – scepticism – but allows that the beliefs generated by particular systems, frameworks, paradigms or societies may count as knowledge-relative-to-that-system, or knowledge-for-that-society. Idealism such as Berkeley's can be seen as a manifestation of this move. Convinced that Locke's moderate realism leads to scepticism, he avoided that conclusion by making the existence of things relative to the appearances of things to the individual. The

pragmatic and coherence conceptions of truth are similarly motivated. The benefits of this sort of move are that the sceptical relativist can make some sense of the fact that people frequently use the
words "true", "know", and "justified" without saying that they are
entirely wrong. A critic might be inclined to repeat Russell's remark
about the advantages of theft over honest toil. But rather than be
content with that, I shall argue that all these views share certain
commitments that are characteristic of scepticism, and so are shared,
for example, with the Cartesian kind of scepticism discussed above.
In particular, if it can be shown that it tacitly employs something like
the K-K or J-J principles then the naturalized or externalist epistemologist will have a reason for rejecting the reasoning which leads to
this conclusion.

Starting from the assumption of variance in standards for hypothesis evaluation, how precisely do we reach the conclusion of relativism or scepticism concerning absolute knowledge? Whether it is the
standards that vary or their application, the variance means that we
can regard scientists operating from within distinct paradigms as
employing different processes of evaluation that may lead them to
differing assessments of the worth of a hypothesis. So let us say that
scientist A evaluates theory T from within paradigm P, while scientist B evaluates T from within a different paradigm Q, and let us call
the processes they employ P* and Q* respectively. According to P* T
is a good theory, worthy of belief, while according to Q* T is a poor
theory and should be discarded. Correspondingly the scientists A and
B form contrary beliefs. Are either of them justified in their beliefs?
Have either of them achieved knowledge? According to the relativist,
the answer is that as long as each has applied the standards and
processes P* and Q* properly, then both will be justified relative to
their own paradigms. The relativist may also want to say, again
relative to their own paradigms, that both A knows that T is a good
theory and B knows that T is a bad theory.

What the relativist will deny, along with the sceptic, is that either
scientist is A-justified or has A-knowledge. The relativist may argue
something like this: say B wants to argue that his judgment amounts
to absolute knowledge and that A's therefore does not. Then B will
need to show that his process of evaluation, Q*, is reliable while A's
process, P*, is not. How may B show this? Since there are no
standards of evaluation outside paradigms, B will have to argue
using the tools of evaluation provided by her paradigm Q. But this
may be seen to give B no advantage, since A may resist B's arguments

from the standpoint of his paradigm P. According to that stand-point A's standards of evaluation P* are more reliable than B's standards Q*.

Therefore, without a paradigm-independent set of standards, we have a stand-off. Neither is able to prove that his or her standards are superior. Hence, argues the relativist, neither is A-justified in their beliefs; neither has A-knowledge.

The nub of this argument may be spelled out in two ways;

(i) First, the relativist may appeal to the symmetry of the situation in which A and B find themselves. Each has what she or he takes to be justification for her or his own views. Neither is able convince the other of the correctness either of their assessment of the theory T or of the reliability of their evaluation processes.

(ii) Secondly, the relativist may suggest that the attempt by B to justify her process of evaluation is bound to end up in circularity. B used the process P* derived from her paradigm P. How might B justify the process P*? B might use P*, but that would be blatantly circular. Perhaps B could use some other process P**. But since all evaluation is paradigm relative, P** must be derived from the paradigm P. Now we may continue this search for justification until either it stops with a process of justification which itself has no justification (perhaps some very general process which makes use of the whole paradigm) – in which case the whole chain of supposed justification is unfounded. Or one in effect justifies the whole paradigm by appeal to the paradigm, in which case, again, the justification is circular.

If these correctly analyze aspects of the relativist's argument, then we may draw out the following beliefs held by the relativist about absolutism:

(a) If one side of an argument has A-justification, then this must ultimately show itself in some apparent asymmetry between their position.

(b) In particular, the asymmetry must be one whereby one disputant is able to show the other, by some set of standards that the other is willing to agree to, that the other is mistaken in some respect.

(c) If a process of evaluation is to A-justify beliefs formed as a result of that process, we must also be able to A-justify that process.

(d) A process of evaluation may not A-justify itself.

Thomas Kuhn

The relativist thinks that absolutism, the claim that A-knowledge and A-justification are possible, is committed to the existence of what we may call an epistemic "Archimedean point", outside science, from which scientific beliefs may be judged and known to be justified or not. Since the relativist thinks that such a point is not available, he thinks that A-knowledge and A-justification are not available either. As Kuhn puts it: "Only a fixed, rigid Archimedean platform could supply a base from which to measure the distance between current belief and true belief ... the Archimedean platform outside of history, outside of time and space, is gone beyond recall ... in its absence comparative evaluation is all there is".[55]

Let us first consider (c). This contains a close relative of the J-J principle. For it says that if S is A-justified in believing that p then S is in a position to A-justify the belief that p is A-justified. Now consider (b). This makes it a requirement of A-justification that it be possible to make A-justification apparent to another disputant whatever paradigm he comes from. First, this implies the same modified J-J principle just mentioned, since if he can make this justification apparent to another disputant then he should be able to make it apparent to himself. Secondly, what sort of process of justification would it be which can be seen to be a justification, whatever one's paradigm? Since one's theoretical commitments are determined by one's paradigm, the justification cannot have a theoretical component. That may leave open the possibility that the justification is one rooted in direct observation. But since Kuhn and the relativist are keen to stress the theory-laden nature of observation, this too is unavailable. That leaves only the possibility of *a priori* justification. Hence the relativist is committed to the thought that the only sort of A-justification there can be is *a priori*. (From which the relativist's conclusion that there is no scientific A-justification follows straightforwardly, since it is clear that there is no *a priori* justification of *a posteriori* beliefs.)

I mentioned that the relativist allows us to use expressions like "knowledge". It may thus be said that A knows that p relative to paradigm P while B knows that not-p relative to paradigm Q. Nonetheless, paradigm-relative knowledge, R-knowledge, is a very different thing from the A-knowledge the sceptic denies exists. As has been stressed, a key component of the traditional concept of (A-)knowledge is that it is factive. S can only know p if it is true that p. The relativist must deny that R-knowledge is factive for any notion of A-truth. If the A-factiveness of knowledge were allowed it follows from the knowledge claims about A and B that both *p* and not-*p* are A-true. So the relativist must either deny the A-factiveness of knowledge – in which case the sense of "knowledge" in which the relativist allows for the possibility knowledge is not the same sense in which the sceptic denies it; or she must consider truth as well to be relative – in which case we have not only epistemological relativism but also metaphysical relativism.

The case as regards justification is a little more subtle. The relativist denies that there is A-justification, just as does the sceptic. But in this case the relativist has furnished us with a concept of R-justification which is not obviously different from our everyday concept. For unlike the case of knowledge just considered, we do allow it to be the case that two individuals may hold beliefs which are mutually contradictory and yet are both justified; justification is not factive. Furthermore, we do think that the assessment of someone's belief as being justified has something to do with its relation to the other beliefs held by an individual. If we substitute "paradigm" for "other beliefs" in the preceding sentence, we get something like a commitment to paradigm-relative justification.

To see why this relativist notion of justification is *not* the same as, or even similar to, our everyday notion of justification, we must consider the role of that notion. "Justified" is one of several positive epithets of the appraisal of epistemic states. In this regard it is in the same basket as "true" (of beliefs) and "knowledgeable" (of people) – and stands in contrast to the terms "unjustified", "false" and "ignorant", which are all negative terms. It is clear why "true" is positive – for instance in seeking to achieve your desires you will be more likely to succeed should your beliefs be A-true than if they are A-false.[56] Indeed it seems that the aim of believing must include the goal of true belief. Given the factiveness of knowledge, knowledge is clearly a good thing if truth is. But why is justification a good thing? In particular, why is it a good thing when it can co-exist with a bad thing – falsity? One way of answering this – an epistemically externalist way – recalls the notion of a process of belief formation. Since true beliefs and knowledge are good things at which belief aims, in producing a false belief, a belief-forming process has gone wrong. Processes can go wrong for different reasons. Imagine a bottle manufacturing process. It may produce flawed bottles because it is a bad process. But it may be a very good process yet still produce flawed bottles if the raw materials are poor. Similarly the explanation for a false belief may be that the process is bad – given good and sufficient evidence it nonetheless produces false beliefs. Or the belief-forming process may be very good but produce a false belief because its inputs were bad, i.e. the process took as its starting point beliefs which were themselves false. According to one externalist account of the concept of justification, its point is to mark the difference between these two cases. A belief may be false yet justified if the belief-forming process is itself reliable: the fault lies with the input beliefs. And we can see why being justified is good, since it marks the fact we have used a reliable belief-forming process. And that is good because by using that process we will tend to produce true output beliefs when given true beliefs as inputs. This is not the only way of explaining the value of justification. On another view, knowledge (rather than merely true belief) is the appropriate goal of belief formation. However, some failures to achieve knowledge are worse than others. In particular, some failures are due to an epistemic failure of responsibility on the part of the believer, while in other cases the believer is epistemically blameless. The role of justification is to mark the latter.

These two explanations of the role of the notion of A-justification, whatever their differences, have this in common. Justification is seen to be a good thing by tying it in some positive fashion to the acknowledged goods of truth and knowledge (where "knowledge" is A-knowledge). However, when we turn our attention to the paradigm-relative notion of justification, we will have difficulty in seeing any such connections. Since the relativist has dispensed with the notion of A-knowledge, it cannot be the link with this state that explains the value of justification. Nor is it any link to the notion of A-truth, for Kuhn insists that later paradigms cannot be said to get closer to the truth. So the paradigm relativist, unlike the naturalized (or externalist) epistemologist, is unable to give any explanation of this kind of the virtue of R-justification. It is important to see why paradigm-relativism is different from the thought that whether a belief is A-justified or not depends on the other beliefs a person has. The reason why the latter are relevant is the same reason as it is relevant to know whether the raw materials of a bottle-manufacturing process are sub-standard or not. If the process produces bad bottles but we see that the inputs are poor, then we have an explanation of the inadequacy of the output that is independent of the quality of the process. Similarly, knowing what a person's other beliefs are will allow us to evaluate the *absolute* reliability of their belief forming-processes independently of the truth of the beliefs they produce, and hence whether, absolutely, the belief is justified.

Some relativists may reject the idea that any of the notions of truth, knowledge or justification have an objective positive nature that requires explanation. Rather, they will say, their use is to allow the scientist's peer group to signify their professional approval of his or her products, while the terms "false", "ignorant" and "unjustified" serve solely to allow expressions of disapproval. They no more signify an attempt at assessing objective qualities than utterances such as "I like this" or "I dislike that". Rather than deal with this at length I shall simply ask whether it is plausible, whether it does justice to the importance we attach to truth, justification and knowledge. Imagine you are shipwrecked on a desert island. You are very hungry. You can find no food. You eventually come across a fungus. You wonder whether it is poisonous or edible. You see that it has a white frill with brown spots. You vaguely recall that toadstools of this description are edible. Then you ask yourself the following questions. "I think it is edible. But do I *know* that it is? Is my recollection *mistaken* or is it *true*?" You then notice that a beetle is eating a fragment of the fungus. Looking at the beetle for a while, you ponder "The beetle hasn't died. So perhaps the fungus isn't poisonous. But then, does the fact that it isn't poisonous for the beetle *justify* thinking that it won't be poisonous for me?" Are your cogitations about the knowledge, truth and justification of your judgments merely idle speculation on whether your absent peer group would endorse them? Or are you asking questions upon which your death by poisoning or avoidance of starvation may depend?

In this section we have looked at relativist arguments that lead from the fact of variance of standards to an incommensurability which rules out the possibility of A-knowledge or A-justification. I have argued that these arguments incorporate some of the same internalist assumptions about the nature of A-knowledge and A-justification that are to be found in traditional scepticism. The relativist wants to avoid overt scepticism and attributes scepticism to the absolutist nature of A-knowledge and A-justification. Hence the relativist avoids overt scepticism by the expedient of dropping the absolutism of these concepts, using instead R-knowledge and R-justification. I have argued that these have little to do with our ordinary notions of knowledge and justification, for they lack the connections those concepts have with *truth*. The connections with the word "truth" might be re-established if we employ a relative notion of truth. But truth is a metaphysical notion, not (exclusively) an epistemological notion – it has to do with the way things are rather than the way we believe things to be. So this approach leads to the thought that the way the world is, is relative to a paradigm – by constructing a paradigm we are constructing a world. It is the naturalness of this move, for the sceptically inclined, along with Kuhn's talk of world change that has give many people the impression that he must have been an extreme relativist or constructivist. As we have seen, there is in fact little in Kuhn to suggest that he did. The multiplicity of worlds thesis is primarily a claim about perceptual experience. Nonetheless, it has some affinity with relativism, since it too is a manifestation of a certain sort of Kantian-cum-empiricist scepticism.

Variance of standards and the possibility of knowledge

If dropping or denying the absolutism of knowledge and justification are wrong ways of avoiding scepticism, what is the correct way? The right way is to reject internalism instead. Externalist epistemology keeps the absolutist component – it allows for A-knowledge and A-justification. Because it denies internalism it denies the K-K and J-J principles. In particular it denies that A-knowledge and A-justification commit one to (a)–(d) listed in the previous section; the relativist is mistaken in thinking that the absolutist must accept these things. It is only when combined with internalism that absolutism is forced to such conclusions.

Consequently, the externalist and the relativist will be in some limited agreement. Some things that the relativist says about R-justification will be accepted by the externalist about A-justification. For example, the relativist denies that if N knows or justifiably believes that p, N should be able to convince a rational opponent that it is true that p or that believing that p is justified. The externalist will agree with this denial. But there will be disagreement too; in particular the reasons given for the denial mentioned will be different. I shall return to this at the end of the section. First, I shall expound what I think an externalist, the naturalized epistemologist in particular, should say about incommensurability due to variance in standards, paradigms in particular.

The naturalized epistemologist does not think that, in general, we can know *a priori* our methods to be reliable. Whether a method, process or heuristic is truth-generating is a question not for philosophical reflection but rather for empirical investigation. This aspect of epistemology – the validation of our belief-forming activities – is a

scientific, naturalistic exercise; hence the name "naturalized" epistemology. Correspondingly many of the processes we use are themselves the products of scientific investigations. If our best scientific
theories tell us that process P* is very likely to give us true beliefs,
then the fact that it is sensible to employ P* will have the nature of a
scientific discovery. Well-confirmed theories will be put to use in
technology, including the construction of instruments such as radio
telescopes, electron microscopes, CAT scanners etc., which will assist
in the generation of further knowledge. Where new beliefs are generated by processes of inference, the credibility of what is inferred will
often depend on the degree to which it coheres with background
knowledge. This is what naturalized epistemology says we should
expect, and it mirrors what Kuhn says about paradigms. If our belief-
forming processes are products of theory then we should expect those
processes to be dominated by our central and best-confirmed theories, the dominant theories of a paradigm.

Let us return to the situation I discussed earlier, where scientist A
uses the standards, P*, of paradigm P to evaluate a theory T, while B
evaluates the same theory employing the standard Q* of paradigm Q.
First, it should be noted that they may agree in their evaluations and
might both be justified in their evaluations – even if P and Q are
competing paradigms. If the paradigms are competing then one or
other will have some key component that is false. But a reliable
process may be constructed on an assumption that is strictly false. If
the false assumption, in the domain of application, is sufficiently
near to the truth to make no difference, then it may be used to generate true results. If the Einsteinian paradigm is right, then
Newtonian physics is false. But it may still be the case that the
numerical results of measurements carried out using Newtonian
mechanics will be accurate if applied to objects of medium size
moving slowly relative to the observer. So both a Newtonian and an
Einsteinian might make the same evaluation of some theory T and
both be justified. (Indeed, in practice, Einsteinians, i.e. any modern
physicist, will typically use Newtonian equations for making calculations where the two theories closely coincide.)[57]

The more interesting case will be, of course, where the two paradigms evaluate the same theory in different ways. It cannot be that
both evaluations are correct – at least one, and perhaps both, will be
in error. Consequently, scientists A and B cannot both have gained
knowledge from the process – one of them has used an unreliable
process. It is quite possible that the unreliability originates with the

falsity of core theories of one of the paradigms. But as long as the other scientist has a reliable process for evaluation, that scientist may well have added to his or her knowledge.

When it comes to justification, matters will be different insofar as both scientists may be justified, even though one or both of them may be wrong. The most obvious way in which this can occur is when the scientists have different evidence. One scientist may simply have more evidence than the other, and the additional evidence may lead him or her to a different conclusion, even when both are reasoning with the same processes. Consequently both can be justified despite the conflicting beliefs. As mentioned, belief-forming processes and methods are frequently built upon scientific theories. We might have a situation where, for good reasons A adopts theory P and for good reasons B adopts conflicting theory Q – perhaps A and B have different evidence, as in the case first mentioned. P generates a reliable method P*, but Q, a justified but false theory, generates an unreliable method Q*. P* and Q* lead A and B to interpret the results of experiments differently. This is plausibly the situation in which Proust and Berthollet or Lavoisier and Priestley found themselves. Are A and B equally justified in their interpretations of the data? A full reply to this question is not possible here. A generous answer would be that since both used methods generated by theories that are themselves justified, both have come to justified conclusions. A less generous answer is that only A is fully justified, since only A's method was reliable. But either way, neither answer would convict B of irrationality; certainly one can understand why B took the route that he did, and even the less generous answer would allow that B is justified in taking himself to be justified in his interpretations.

The conclusion is then that although there is no need for a notion of paradigm-relative knowledge, there may be some room for a very mild form paradigm-relative justification. But the latter is not because paradigms are the ultimate sources of justification but because adherents of different paradigms have different evidence at their disposal, and so are justifiably inclined to believe different theories, to adopt different methods and to construe data differently. Here "relative" does not contrast with "absolute". Given appropriate information about evidence and so forth, it will be an absolute, paradigm-independent matter whether a scientist is justified. The role of the paradigm is just that it encapsulates the evidence and belief-forming processes scientists have at their disposal.

So far in this discussion I have been using "paradigm" in the broad sense of "disciplinary matrix". I shall also make some remarks about the narrower sense of paradigm, where the paradigm is an exemplar, typically visual, used in the training of pattern recognition and in the learning of similarity relations. In Chapter 3 I described how Kuhn's remarks could best be understood in a connectionist way. According to connectionism, training tunes the brain or the human system more generally to be responsive to certain kinds of input. Recalling Kuhn's example, the aim might be to get the subject to recognize wolves, differentiating them from dogs. Early on in the training the subject's detection system is likely to be an unreliable instrument, giving incorrect outputs – "wolf" for some dogs, or "dog" for some wolves. But with regular retuning and long enough training the system achieves a much higher degree of accuracy, perhaps 100 per cent for certain kinds of case. What has been produced is a reliable instrument for the visual recognition of wolves. According to the naturalized epistemologist, a reliable instrument of this kind will give its possessor knowledge of whether a creature is a wolf or not. The connectionist story tells us how perception works, naturalized epistemology tells us (what we knew all along) that perception can give us knowledge.[58]

The foregoing remarks are intended to illustrate what the naturalized epistemologist might say about a difference in theory evaluation due to variance in standards, and also about the recognition of similarities by a perceptual system. At the beginning of this section I said that the relativist and the externalist could agree on some things, but would disagree on others. For example, they will agree that a satisfactory notion of justification need not satisfy conditions (a)–(d) in the last section. The relativist thinks that A-justification does require them, but I think we can now see that the externalist notion of A-justification does not. The relativist had two arguments. The first appealed to the thought that if A has A-knowledge or A-justification and B has not, then this epistemic asymmetry should show itself in some more apparent asymmetry, one which both are able to see directly. Put another way, A ought to be able to convince B that A's evaluation is better justified than B's. The naturalized epistemologist will see no reason for this being true. There will be an asymmetry – A's process of evaluation will be reliable where B's process is not. But the naturalized epistemologist does not agree that this asymmetry is one that must be apparent to both A and B. Since the methods of appraisal in a mature science will be built on theories that are themselves built on, or appraised relative to, other theories and so on, it is quite unlikely that scientists will be able to provide a justification of their methods that can be appreciated by someone operating in an alien paradigm. Since what is required for knowledge is the actual reliability of the appraisal process, that does not matter. In the long run it may be that we all share some common grounds, low-level theory related to innate epistemic capacities, simple perceptual observations etc. But there is no reason to think that any scientist can or ought to be able to justify his or her activities in these terms. Take an example of someone whose methods of justification are clearly unjustified. This individual has at the core of his paradigm the conviction that the Earth is flat or that the Moon is made of green cheese. Any observation or theory inconsistent with this is explained away by ad hoc hypotheses. Given enough ingenuity it might be possible to keep this up for some while (and even when the going gets tough this individual may just blame his own intellectual shortcomings for failing to explain away anomalies). There seems to be no reason to expect that we should be able to convince this person that his paradigm and evaluations are unjustified. Of course, we may dismiss such an individual as mad. But just as it may be difficult to convince the madman that he is mad, it may be difficult to convince a sophisticated scientist that he is unjustified. More importantly, the typical case is not of an opponent who is unjustified, but rather an opponent who has a some high degree of justification but is nonetheless mistaken. No argument you may be able to think of may be relevant to the case. You will just have to hope that in the future the mistake leads to anomalies that will convince your antagonist on his own terms of his error.

The situation just described is one of which the relativist is more aware than most. Relativists regard it as evidence *for* relativism. The thrust of my remarks is that the externalist epistemologist may equally acknowledge this fact – and explain it – while holding on to absolutism. I said that the relativist evades scepticism by dropping absolutism. But that is not the only motivation for relativism. The relativist's intended contrast is with certain Old Rationalist views. Old Rationalists tended to see a special role for philosophy in the epistemology of science: delineating the scientific method, drawing up an inductive logic, or providing a deductivist account of theory-preference. Armed with the products of philosophical epistemology it should then be possible to distinguish good science from bad, knowledge from false belief, and justified, well-grounded theories from irrational ones. Old Rationalist epistemology *does* accept the possibility of an Archimedean point; anyone in possession of the proper philosophical account of scientific method (or inductive logic, etc.) will be at that point. Lakatos' methodology of scientific research programmes sets out to describe the criteria by which a scientist's development of a theory or her choice among competing research programmes is rational. A recent debate between Larry Laudan and John Worrall also illustrates the Old Rationalist belief in the Archimedean point.[59] Laudan takes on board Kuhn's view that methodological beliefs in science have changed. He provides what he calls a "reticulated" model of scientific change whereby beliefs at three levels – theories, methods and aims – may change in response to one another. Worrall's response to this is Old Rationalist through and through:

> Assuming that changes in real, "implicit" methodology are at issue, then Laudan's 'reticulated model', as it stands, collapses into relativism. If no principles of evaluation stay fixed, then there is no 'objective viewpoint' from which we can show that progress has occurred and we can say only that progress has occurred *relative to the standards that we happen to accept now*. However this may be dressed up, it is relativism. Without fixed standards, no amount of 'mutual adjustment . . . among all three levels of scientific commitment' can avoid it.[60]

Worrall's fixed standards comprise an Archimedean point. It is not surprising that if anti-relativists such as Worrall think that abandoning the Archimedean point is to commit oneself to relativism, then the relativists should think so too. But both are wrong. It is not absolutism that requires the Archimedean point but absolutism plus internalism. Give up internalism and one can happily reject the possibility of an Archimedean point, as does the relativist, but without giving up absolutism. Naturalized epistemology makes this perfectly clear. What someone knows, whether their beliefs are justified, whether science has made progress, are empirical matters like any other. Being in a position to evaluate them does not require a "fixed viewpoint" – it requires only enough evidence and a method of enquiry sufficiently reliable for this purpose. Consider a doctor who is concerned to know whether her patient is getting better or deteriorating. Knowing the answer does not require some special, fixed method of medical prognosis. It requires only that we now have techniques (scans, blood tests and so on) that give a reliable answer. We did not always have such techniques and no doubt we shall develop better ones. Not just any method will do, but that does not mean that there is only one unchanging method that will. Similarly, knowing whether Lavoisier's chemistry was an advance on Priestley's does not require an Archimedean point. It requires only that we have sufficient evidence and a reliable way of measuring progress in eighteenth-century chemistry. The two cases are not merely analogous; they are both instances of getting to know the answer of a question of contingent fact, using an appropriate method. The question about progress does have one special feature which is that it is difficult to answer with regard to current science. For example, we may ask whether current theories are an advance on their predecessors in the sense of giving us knowledge that the latter did not. Being entitled to assert that they do requires knowing that current theories give us new knowledge. In rejecting the K-K principle I pointed out that in general it is easier to know that *p* than to know that one knows that *p*. If current theories do give us new knowledge, it is likely to be a difficult intellectual achievement; it will be an even more difficult intellectual task to know that we have this knowledge, a task that we are unlikely to be equipped for. So it is quite likely that we are not in a position to tell whether recent theoretical changes constitute additions to knowledge. This admission is a consequence of giving up on the Archimedean point. But it is hardly a painful admission, and there is no reason to suppose that we will never acquire the resources required to answer the question at issue.

Conclusion

The central element of Kuhn's epistemological outlook is his neutralism about truth. Methodologically this is acceptable. But there is no justification for elevating it into a metaphysical principle. Kuhn's arguments against the concept of absolute truth are confused. In mitigation it must be said that this is a confusion he shared with his positivist predecessors. But that is yet additional evidence for my contentions that Kuhn retained many positivist and empiricist assumptions and that their retention is in large part responsible for some of the more radical of his pronouncements. The attack on truth (which was also an attack on realism) is tantamount to a deep, metaphysical form of scepticism or, in the context of the later conceptions of incommensurability and lexicon-relative truth, relativism. This metaphysical relativism is also a consequence of Barnes' finitism.

While Kuhn's talk of world changes gives the impression of endorsing metaphysical relativism, I do not think that was his intention. Nor do I think, for that matter, that this was Barnes' aim either. That those consequences are unintended and that they in turn lead to contradiction are reasons for retreating from the relativization of truth and from finitism.

Just as a commitment to methodological neutralism does not require rejecting or relativizing the concept of truth, retention of that concept is consistent with Kuhn's Darwinian account of progress. Even so, Kuhn's argument that the progress of science is not towards the truth needs strengthening, since the Darwinian analogy is not enough to rule out the possibility that our scientific beliefs are getting closer to the truth. The historical argument against convergent realism, that examination of past scientific beliefs show little sign of increasing verisimilitude, is just the sort of strengthening required. But it should be borne in mind that there could still be increasing knowledge even if there is no overall increasing verisimilitude.

Kuhn of course rejects the views of science as achieving increased verisimilitude and as accumulating knowledge. He is a sceptic with regard to A-knowledge; there is no knowledge, where knowledge is taken to entail A-truth. To a considerable extent knowledge and justified belief drop right out of the picture, since for Kuhn the object of epistemic assessment is not belief but belief change.[61] Kuhn does not say *why* we should not assess the beliefs themselves, but in the context of the aim of science being puzzle-solving rather than truth it is plausible to suggest that Kuhnian scientists should not really *believe* their preferred theories. A Kuhnian scientist, like van Fraassen's constructive empiricist, is instrumentalist in outlook. However, to the extent that scientists do believe their theories, or at least the longest established and best confirmed of their theories, it is tempting to assess their beliefs in epistemically relative terms. The temptation does not escape Kuhn, who talks of choosing between rival bodies of knowledge and of knowledge production where what is produced is false.[62] The relativist rightly rejects the idea of a fixed Archimedean point from which all beliefs may be judged for their justification and assessed as potential contributions to knowledge. For Kuhn the issue is the rationality of a particular move in the history of science, say a revolutionary choice of new paradigm, and that may be assessed from a moving platform consisting of the beliefs shared between the potential paradigms on offer.[63]

I have argued, however, that the possibility of A-knowledge does not require a fixed Archimedean point or platform. But its replacement should not be the viewpoint of the subject or the professional group. Certainly the assessment of others' judgments and choices of theory requires knowing their background beliefs and may involve hermeneutic interpretation. But externalist epistemology says that the assessment will also need to draw on information that may not have been available to the subject. In particular we need to be able to assess the reliability of the belief-forming process used. That is an assessment of a contingent matter of fact. Just as when we judge any contingent matter of fact, we must use whatever techniques and knowledge *we* have to hand. Our judgment is fallible but need not always be mistaken. When assessing the justification of S's judgment that *p* we replace the fixed Archimedean point not by S's paradigm; nor do we replace it merely by our own point of view on the subject matter *p*. Instead we replace it by our point of view on the relationship between S's psychology, S's evidence, S's environment as well as the subject matter *p*.

The externalist response to scepticism and relativism is buttressed by the naturalistic approach to epistemology. In this view knowledge is the result of exercising cognitive capacities that have in some cases evolved over millions of years. There is selective pressure for capacities that generate mental states that represent the world accurately, as exemplified by Kuhn's case where there is pressure to be able to represent wolves differently from dogs. Learning through paradigms is a local instance of the same general phenomenon. Training with exemplars works thanks to the evolution of a neural network, as described by connectionist neuroscience. Kuhn's appeal to this way of knowing is a small piece of naturalized epistemology. While Kuhn is resistant to naturalism, this I suspect is an empiricist legacy.[64] Had he thrown off this residual empiricism, a thoroughly naturalistic Kuhnian picture could have emerged. Naturalized epistemology allows for the possibility of knowledge and so rules out philosophical scepticism. But it does not demand that the possibility be actualized – that question is for empirical investigation to answer. A naturalized Kuhnianism might well be sceptical in a scientific fashion. Knowledge of similarities to exemplars might be admitted while knowledge of the truth of theories might be denied on the grounds that as a matter of fact the motor of theory choice is puzzle-solving ability, and that this is too poorly correlated with the truth to be a source of knowledge.

Chapter 7

Kuhn's legacy

Social science after Kuhn

The Structure of Scientific Revolutions had little direct influence on the functioning of the natural sciences but its impact on social science was enormous. This impact had two aspects; the first was a change in the social sciences' self-perception, the second was a suggestion of a new role and subject matter for the social sciences. Kuhn's cyclical pattern provided a template whereby the histories of the social sciences, almost entirely ignored in *Structure* itself, could be described. The significance of this was not so much the furthering of historical research; rather it was the provision of a tool for validating social sciences as true sciences. Two important and obvious characteristics of the leading natural sciences – physics, chemistry and biology (taken to include the biomedical and agricultural sciences) – were absent from the social sciences and thus seemed to mark the former as genuine sciences and the latter as impostors. The first characteristic was that the natural sciences had vast and powerful technological applications. While many social scientists did believe in the possibility and value of "social engineering" and other applications of social research in areas of social policy, they found little in the way of unequivocal success. Kuhn's description of science had nothing to say at all about the application of science. He scarcely mentioned the thought that technology might be a source of problems and puzzles for scientific research; still less did he regard the possibility of technological application as any kind of sign of sciencehood. Rather, the mark of being a science was that a discipline should have a certain sort of internal dynamic – puzzle-solving research governed by a paradigm, interrupted by crises and revolutions – a dynamic that could be possessed by a discipline however bereft of associated technology.

The fact that Kuhn has no place for the truth of theories in his description and explanation of this dynamic, plus the central place given to revolutions that reject or revise existing beliefs, is a further reason why the social sciences should not, after all, be faulted in comparison to the established natural sciences. The latter seemed to have impressive histories populated by heroic figures, histories that had generated great tracts of accepted knowledge and whose continuing progress is governed by well-established methodologies. This was the second apparent sign of sciencehood. The social sciences, by contrast, were young, had few acknowledged heroes, had produced little in the way of knowledge that was widely accepted, and seemed exercised less by the need to make progress than the desire to engage in methodological disputes. *The Structure of Scientific Revolutions* rendered the contrast less acute. On the one hand, it undermined the positive image the contrast ascribed to the natural sciences – knowledge is not cumulative, the difference between the heroes and those they vanquished is not that between truth and falsehood or rationality and irrationality; scientific progress as traditionally understood is an illusion; and there are no universal methodologies, just those licensed by particular paradigms. In general, the traditional view of science as progressing inevitably towards the truth (much of which we know today) – the so-called "Whig" history of science – can be dismissed as a myth, one generated by the very nature of the training that is central to maintaining the scientific enterprise. On the other hand, the unflattering contrast afforded to the social sciences need not refute their claims to sciencehood. Even if methodological and other disputes do predominate in a field, that is compatible with its being in a state of revolution or, more plausibly, in a state of immature science, which, as Kuhn says, is no less truly scientific for all that.

Kuhn partially endorsed this assimilation of the natural and human sciences. "I'm aware," he says, "of no principle that bars the possibility that one or another part of some human science might find a paradigm capable of supporting normal, puzzle-solving research," adding: "Very probably the transition I'm suggesting is already underway in some current specialties within the human sciences. My impression is that in parts of economics and psychology, the case might already be made."[1] He remarks that perceived differences between the natural and human sciences depend on a "relatively standard, quasi-positivist, empiricist account of natural science, just the image that I hoped to set aside".[2] Social scientists in particular are apt to think of the natural scientist's confrontation with natural reality as essentially unproblematic in a way that does not hold for themselves. The social sciences are necessarily hermeneutic – they do not seek to describe the laws of nature but instead to understand or interpret human behaviour. It is the need for interpretation, the finding of meaning, that is absent from the natural sciences. Kuhn thinks that the social scientists misconceive the difference. The world with which the natural scientist engages is not fixed and does not supply its own natural and unchanging categories for the scientist to employ. The world, the phenomenal world, is dependent on the variability of paradigms, which furnish the concepts through which the natural scientist sees the world. To carry out science within a paradigm is to engage in an interpretative exercise. Kuhn is nevertheless keen to point out that there are indeed some significant differences, even if they are not where we thought they were. Notwithstanding the fact that practising natural science involves interpretation, its *aim* is not interpretative. Scientists engaged in puzzle-solving normal science do not think of themselves as interpreting (as opposed to inferring), while social scientists are actively engaged in looking for new, deeper interpretations.[3]

Even this difference may be overstated. If one thinks of there being genuine explanations and causes of behaviour then one might think that interpretation is not different from inference but is instead an instance of it. It is true that there are aspects to understanding that are not like the confirmation of a hypothesis: one can empathize with a subject, one can simulate their thought processes as a way of seeing why they do what they do. Nothing like this exists in chemistry, physics or biology. On the other hand, once we see (with Kuhn's help) that there is no single, special method that is common to the natural sciences we may be inclined to see understanding ("verstehen") not as a mark of radical difference but instead as just another tool in puzzle-solving, one appropriate for (some) social sciences.[4]

The social study of science

The Structure of Scientific Revolutions had a more specific but nonetheless widely felt influence on the social sciences. Before Kuhn the history of science might have been scholarly but was never theoretically interesting. The motor of scientific change was tacitly understood to be the scientific method or scientific rationality which according to Whig history would inevitably drive our predecessors to our current state of knowledge. The only role for the student of science was to reveal the particular details of this inescapable trajectory. After Kuhn the study of science looked potentially far more interesting. As a work of theory itself *Structure* showed that theoretical and not merely detailed empirical research was possible in the history of science. The fact that Kuhn drew philosophical lessons from his historical and social account of science made this new approach especially attractive. The emphasis on science as an institution suggested that it would be suitable for the sort of study afforded other institutions such as religion or the state. The jettisoning of the Whiggish, Old Rationalist motor for scientific development and its replacement by an evolutionary account opened up the possibility of more particular, local explanations of change in place of the global "scientific method". In particular many were impressed by the hints Kuhn gave that explanations of revolutionary change need not be entirely intra-scientific and that political, social, religious and personal motivations might play a much larger part in the history of science than had previously been admitted. For some the effect was to reinvigorate Mannheimian and Marxist history of science which in the hands of Bernal, Sarton and Needham had struggled but not quite succeeded in freeing itself from a traditional perspective. For others, such as Latour and Woolgar, a stronger social constructivism that traces its origins via Kuhn back to Fleck was exemplified in their study of the details of how what is taken to be a scientific fact becomes established as such. If the "facts" are indeed socially constructed then radical concerns with the nature of society will generate radical critiques of science itself: hence Knorr-Cetina's *The Manufacture of Knowledge* leads to Harding's *The Science Question in Feminism*.

The Structure of Scientific Revolutions may have initiated this revolution but it scarcely operated as a paradigm. Kuhn and his ideas played little part in these developments. Perhaps the school in the science studies world closest to Kuhn's own thinking has been the Science Studies Unit at the University of Edinburgh and its Strong Programme. As we have seen, Bloor and Barnes adopted Kuhn's neutralism as a methodological precept. Their study of the social conditions of the production of scientific belief could be understood as constructivist, and was originally regarded as such. But it may also be understood in a much more naturalistic fashion, as is suggested by another precept of the Programme, that the social study of science itself must be regarded as a science and so not outside the scope of its own conclusions. The Strong Programme combines both of Kuhn's influences on the social science: the sociology of scientific knowledge is both able to study science and to declare itself a science too. Shorn of its finitism and so the tendencies to a metaphysical relativism and social constructivism (taken more from a particular interpretation of Wittgenstein than from Kuhn), a naturalistic Strong Programme is the appropriate replacement for the *a priori* accounts of scientific rationality and method found in the Old Rationalism and which Kuhn rightly encourages us to eschew.

Since Kuhn was absent in all but his oft-mentioned name in post-Kuhnian science studies, he felt it necessary to repudiate much of what passed as Kuhnianism. His last published paper, "The Trouble with Historical Philosophy of Science" (1992), is devoted to distancing himself from the Strong Programme, which earlier he had accused of post-modern excesses.[5] Kuhn had a caricature of the Strong Programme in mind whereby all theory choice is down to politics and power, and that which gets called "knowledge" are just the beliefs of the winners. Whether or not the caricature is accurate the important thing is that Kuhn wanted to distance himself from a view of science that has no room for the role of evidence and reason. Bloor, on the other hand, is ready to defend Kuhn against charges of irrationalism: "Perhaps the most shameful of all misunderstandings was the idea that, for Kuhn, science has no significant contact with an independent reality". Bloor goes on to point out that, as we have seen, the world change thesis does not imply idealism. Even so, Kuhn continued to be disparaging about the idea of "an independent reality".[6] "I am not suggesting, let me emphasize, that there is a reality which science fails to get at. My point is rather that no sense can be made of the notion of reality as it has ordinarily functioned in philosophy of science." He goes on to explain: "what replaces the one big mind-independent world about which scientists were once said to discover the truth is the variety of niches within which the practitioners of these various specialities practice their trade . . . unlike the so-called external world, they are not independent of mind and culture, and they do not sum to a single coherent whole of which we and the practitioners of all individual scientific specialities are inhabitants".[7] In Kuhn's Kantian picture all that can be discussed are the evidence and the states of the particular scientific specialities.

Social knowing

Bloor says that Kuhn espoused the social construction not of reality but of knowledge.[8] If this is right then what is meant by "knowledge" is not A-knowledge, since Kuhn is a sceptic about A-knowledge. "Knowledge" here must be something weaker, either R-knowledge or accepted scientific belief, that which is *taken* for knowledge. In fact it is this, the social generation of accepted belief, that is the focus of the work of Latour, Woolgar, Shapin, Knorr-Cetina and other social constructivists, even if they give it more impressive epistemologically or metaphysically laden names such as the social construction of "knowledge", "truth" or "fact".

Kuhn rightly reminds us, as did his philosophical antagonist Popper, that in the scientific context what we call knowledge is essentially social. Social knowing is what we talk about when we think of "the growth of scientific knowledge" or when we say "it is now known that smoking is a cause of cancer". The tendency among the Old Rationalists, one largely shared by contemporary philosophers, was to see epistemology in individualistic terms. Hence even social knowledge is reducible to individual knowledge – cases of social knowing are instances of facts being known widely by individuals. This is a mistake. For much of what is socially known is not widely known at all. A scientist might say "it is now known that some ceramics show superconduction at temperatures above $-200°C$". She may be right, but surely that fact is known by a relatively small proportion of the population. Nor is it sufficient for social knowledge that someone knows something. A reclusive scientist may make a discovery which he keeps secret to his death. He may have known something new, but that is no addition to scientific knowledge.

Indeed, I suspect that being known by some individual is not even necessary for social knowledge. For instance, a state defence establishment may have knowledge of how to build an atomic bomb without any individual having that knowledge. In that case the individuals may posses parts of the social knowledge, but there may be other cases where even that is not true. Consider the following scenario: a scientist works in some obscure field of little interest to his contemporaries. The work is of high quality and generates new (personal) knowledge for the scientist. He publishes his research in a highly reputable journal – the standard repository of research in that general area. Other scientists note that the paper is published but do not bother to read it. This scientist dies (and so, coincidentally, do any referees who read his paper for the journal). Shortly after, new discoveries suddenly make his work significant and remembering that he published this work (or by making searches in the bibliographies and indexes of this science), scientists now read his paper for the first time. I claim it is natural to describe the case as follows. On publishing his paper the scientist makes a contribution to scientific knowledge, although he was the only individual who had this as personal knowledge too. When he dies the contribution to scientific knowledge remains, even though no one has personal knowledge of it. Later more individuals possess personal knowledge of the discovery when they first read the paper.

Thus social knowledge is not reducible to individual knowledge. Unfortunately Kuhn and the social constructivists, because of their scepticism, failed to give a positive account of what social knowledge is and how it differs from social belief. For Kuhn and the social constructivists alike the task is the same, explaining why a group of scientists believed at some time what they did believe. However, in the light of naturalized epistemology there is more that can be said. For we may wish to distinguish those social processes by which social knowledge is produced from those which fail to produce knowledge even if they succeed in producing widespread belief. The difference between propaganda and careful academic research may be not only social but also epistemic. The new role for sociology in naturalized social epistemology is to investigate the reliability of mechanisms of social belief production and to explain how that reliability originated and is maintained. Take for example the importance of testimony. Many things we read or are told come from unreliable sources and cannot count as knowledge. But others are reliable, and one of the reasons why a source may be reliable is that social selection or social pressure ensures that it is reliable. Sensational newspapers include stories that no one ought to believe, but at the other extreme one can trust the financial press to give accurate reports of yesterday's share prices – if it failed to do so, a paper's reputation and revenues would crumble. The latter is a social explanation of why journals such as the *Financial Times* are respected for their reliability, and why, therefore, they are repositories of knowledge as opposed to purveyors of mere belief or gossip.

Something similar may be said about the social organization of science. Kuhn emphasizes the uniformity of scientific education and the pressures that enforce conformity during normal science. To some, Popper for example, this may seem like a conservatism that is inimical to the unfettered search for novel truth. But as Kuhn points out, some degree of dogma and conservatism are essential if science is to make progress in normal science, by building on achievements rather than always challenging them. Furthermore, if the paradigm at the heart of normal science is true or close to the truth, then such conservatism will itself be conducive to the truth. Conservatism will both promote the acquisition of new true beliefs by providing a firm basis for research, and it will also avoid error in protecting the paradigm from replacement by a false theory. At the same time it is true that if the social forces are excessively conservative, with no room for challenging accepted belief, it is as likely that a false paradigm will be entrenched as a true one. So one needs to strike a balance between on the one hand enough liberalism and sufficient willingness to listen to criticism, in order that falsehood does not become firmly established, and on the other hand enough conservatism to provide a stable platform for the development of new knowledge. Kuhn's model provides this too: a paradigm is not adhered to willy nilly, but may be rejected in the light of anomalies. So, prima facie at least, Kuhn's model of development is consistent with the claim that the social organization of science is well suited to knowledge production.

Kuhn's explanation of the behaviour of individual scientists is primarily in psychological terms that do not seem to reflect the search for objective, absolute knowledge. The explanations of puzzle choice and puzzle-solution are in terms of a psychological sense of similarity to an exemplar inculcated by a rigid education. During normal science, success in puzzle-solving is not supposed to be regarded as confirming the paradigm theory. Correspondingly, crisis is depicted as a state of anxiety rather than one of acceptance of the rational refutation of the paradigm by evidence. A scientist's preference in the choice of new paradigm seems even less governed by the goal of truth than is puzzle-solving in normal science. Does not the (alleged) fact that scientists are not truth-oriented demonstrate that science is not organized so as to produce knowledge? Kuhn is not entirely just as regards the motivations and aims of scientists, but even if he were, that would not prove that science as an institution cannot produce social knowledge. The psychological states ascribed to the individual scientists may be just what is required for the social system to ensure that its favoured hypotheses have a good chance of being true. For example, the driving motivation for scientists could be professional success; but if professional success is measured by criteria that correlate well with truth, then this motivation will bring about truth-tropic behaviour.

Nonetheless, there are features of at least the simple Kuhnian model that would tell against the social production of knowledge. First, if normal science is always conservative and non-revisionary and if revolutions are always major changes, then there would seem to be no mechanism for correcting minor errors. Secondly, if revolutions were to involve total rejection of earlier paradigms, there would be no expectation of the long-term survival of true beliefs. Thirdly, if revolutions are decided by extra-scientific factors, then there would be little reason to think that even an erroneous paradigm theory would get replaced by a true one. Fourthly, Kuhn's rejection of the possibility of non-revisionary revolutions seems to limit the scope of what could be known. So a social system correctly described by the simple model would be unlikely to be one that is well adjusted for providing knowledge. For each of these theses I have argued that either Kuhn does not hold that thesis or that he ought not to. We have seen that his understanding of progress through revolutions does provide for continuity, with the preservation of content from earlier paradigms. On inspection it appears that scientific change is not polarized between extremes of normal and revolutionary activity, but instead there is a continuum between them. There can be, after all, mini-revolutions; revolutions can take place not only on the grand scale but also at the detailed level of a small sub-discipline or micro-speciality. And I have also suggested that significant revolutions may occur in a fashion that is non-revisionary. Normal science is not rigidly conservative; criticism is a perennial feature of scientific life. Peer review of journal submissions serves to eliminate obvious falsehood. But it would be a mistake to think that the pages of journals are filled only with indisputably correct solutions to puzzles and minor problems. On the contrary, they are characterized by lively debate and disagreement.

Kuhn's work has certainly encouraged the sociology of science, but that field has tended merely to emphasize relativism as a methodological precept. Rather, in a naturalistic spirit, sociology might also ask whether or not the institutions and social organization of science are conducive to social knowledge production.

Conservative Kuhn

Bloor characterizes Kuhn as a conservative thinker. There are a number of senses, not all directly related to one another, in which

Kuhn might be thought of as conservative. For Bloor, Kuhn is a conservative in the Mannheimian sense according to which the conservative stresses the importance of tradition.[9] This is certainly a fruitful approach to understanding Kuhn, especially in contrast to the Old Rationalists who exemplified what Mannheim calls the "natural law ideology". Natural law ideologists, the inheritors of the Enlightenment, emphasize reason conceived of as the application of general and sempiternal rules of judgment; the conservative appeals to history and local factors.[10] Translated into the philosophy of science, the natural law approach looks to an *a priori* and perfectly general scientific method which will lead belief towards the truth. Kuhn's conservative alternative depicts science as made up of largely disunited practices that progress according to the rule-less impetus of their current paradigms, adjusted not with a view to representing the facts but instead with the aim of making the best local improvement to the existing tradition.

Kuhn has recently been accused of conservatism on a second, more explicitly political count, by Steve Fuller.[11] Fuller's criticism has several strands. *The Structure of Scientific Revolutions* must be seen, he says, against the educational background that caused Kuhn to write it. In the early 1950s Kuhn taught the general education in science curriculum at Harvard. This programme had been devised by Harvard's president, James Conant, as a way of informing students, humanities students in particular, about the nature of science and its history. Conant was a central figure in the liberal-conservative establishment, a supporter of Big Science, who left Harvard to become the first US ambassador in West Germany. The motivation behind the programme was to build in the minds of America's future leaders a particular image of science: science as an autonomous institution. Science needed to be protected from governmental interference, and so the elites needed to be reminded that the dynamic of science is generated internally not externally. Science also needed to be defended from criticism directed at the uses of science such as the application of physics in the design of nuclear weapons. So the second reason to reinforce the autonomy of science is directed towards science's public legitimation. Science needed to be divorced in the public mind from technology. Kuhn's image of science certainly fulfils these requirements. Science generates its own puzzles; good science occurs when it sets its own agenda. And what drives science is the desire to solve these puzzles, not the need to produce inventions however beneficial or deadly. If a puzzle-solution happens to be of use in some way then putting it to that use is a part not of science but of technology.[12] A second strand in Fuller's criticism concerns what he calls the "double truth" doctrine, which is really a "double doctrine" doctrine. The general idea is that the masses are not given the same message as the elites if to do so would be socially destabilizing. Thirdly, far from encouraging critical attitudes towards science (which he repudiated) Kuhn's work blunted criticism. In general, the process of crisis, revolution, replacement of an old paradigm by a new, revisionary paradigm, shows that science has its own mechanisms for self-criticism, change, and renewal. Hence external criticism is unnecessary. In particular, philosophical criticism is blunted, first, because the emphasis laid on dogma in normal science is highly conservative especially when compared to Popper's critical rationalism, and secondly, because the empirical, historical attitude towards science removes the normative element from the philosophy of science. At the same time *Structure* encouraged social theorists to see themselves as scientists rather than as critics of science.

It is not clear that *Structure* had quite the conservative impact Fuller suggests. Is it really true that there would have been more and better criticism of science had it not been for that book? The fact that it has been used for some conservative legitimatory purposes (for example by Francis Fukuyama) does not really show that Kuhn's work significantly furthered those purposes, let alone that Kuhn shared them. It is the fate of iconic texts to be pressed into the service of all sorts of causes, and in Kuhn's case it is clear that citations of him by science's critics are not meagre. Fuller extracts the double truth doctrine from Kuhn's remarks concerning the teaching of the of the history of science to science students. The latter receive a cleaned-up, Whiggish kind of history, unlike the story Kuhn tells us of crises, revolutions, incommensurability, and so on. On that reckoning *Structure* contains the elite message. But that is incompatible with the role ascribed to the book in Fuller's other criticisms, as conveying its message to the public, which will include the very scientists from whom the elite message must be hidden. In any case, it is not clear that Kuhn thinks that scientists should not be given a warts and all history of science if they happen to be studying the history of science for its own sake. If history of science is to be used at all in teaching scientists to be *scientists* the tidied up image is more helpful. As a matter of fact, as Kuhn points out, the historical element in science teaching is thin, unlike in the humanities, and Kuhn's remarks on the topic are scarcely central to his account; even less do they amount to a plank in a political agenda. While Fuller's treatment of Kuhn and his context is full of erudition, telling parallels, and insightful suggestions, ultimately his case against Kuhn rests on association rather than documentary proof; the evidence is circumstantial rather than concrete. And even if Fuller's account of the true nature of Kuhn's work were correct, that would not obviously impact on our assessment of his philosophy *as* philosophy or his history *as* history.

Kuhn and the legacy of empiricism

A third sense in which Kuhn could be characterized as conservative is represented by the central theme of this book: that despite radical appearances to the contrary, Kuhn's unstated presuppositions are really not so different from those of the positivists, empiricists and Cartesians he took himself to be criticizing. In the preface I remarked that Kuhn started his professional career as an historian but finished it as a philosopher. Kuhn's contributions to philosophy should be seen as consequences of his attempts to treat the history of science in a theoretical manner. He stated that revolutions are often brought about by those who come to a field from outside, and his work would certainly seem to illustrate that. As we have seen, scientific revolutions are not root-and-branch revisions of belief, and as I emphasized at the outset Kuhn's own revolution retained much from the logical empiricism against which he was reacting. I likened Kuhn to Copernicus who, while striking the first serious blow that brought down the Ptolemaic-Aristotelian world-view, was also irrevocably steeped in that way of thinking. Kuhn himself said of Copernicus's revolution that it was like the midpoint of a bend on a road: from one perspective it is the last point of one stretch of road and from another it is the beginning of the next. Similarly Kuhn can both be seen as among the last of the empiricists and also be regarded as the first of empiricism's successors.[13]

Logical empiricism is no longer the force in the philosophy of science it was when Kuhn was writing. Kuhn's contribution to that end was primarily his rejection of the independence assumption in the observational basis – the idea that the nature of perceptual experience is independent of an individual's mental history. But the philosophical advances that led more surely to the demise of empiricism were those that took place after and independently of the publication of *The Structure of Scientific Revolutions*. Indeed, Kuhn either ignored or explicitly rejected these developments. The developments I have in mind are: (a) advances in the theory of reference, in particular externalist, non-intensionalist and causal approaches; and (b) advances in epistemology, namely naturalized epistemology and externalism. It is no coincidence that "externalism" is mentioned in both sets of developments. Externalism about a mental state says that the nature of that state may depend on features that are external to the individual; internalism denies this. The mental state may be that of thinking about something, in which case the internalism–externalism debate is one about reference. Or if the mental state in question is that of knowing, then the debate is epistemological. Just as Copernicus' heliocentrism when combined with Ptolemaic mechanisms and Aristotelian physics led to an unstable combination of beliefs, so too did Kuhn's only partial rejection of empiricism. It is the retention of an empiricist brand of thick intensionalism that supports his radical-seeming claims about incommensurability. Had he rejected a view that he shared with the likes of Carnap and instead adopted a non-empiricist view such as Kripke's then the philosophical issue of incommensurability would never have arisen. On the epistemological side, concerns about the possibility of knowledge encouraged at best a neutralism with strong relativist tendencies. Such concerns arise from the perceived need for a neutral standpoint in order for knowledge to be possible – an Archimedean standpoint that the doctrine of paradigms rules out. Such a perception stems from an internalist epistemology which demands that for knowledge the reliability of the process that yields it be apparent to the subject. Here too externalism would have halted the drift to relativism by denying this requirement on knowledge, thereby removing the need for a neutral, paradigm-independent standpoint that Kuhn recognized to be unfulfillable.

To argue that Kuhn did not complete the revolution against empiricism is not to criticize or underestimate his achievement. His influence has been immense. It is perhaps ironic then that one of Kuhn's most original and potentially fruitful contributions to thinking about science was largely overlooked or misinterpreted. The term "paradigm" has become an empty cliché and perhaps because of this the underlying idea of perceiving similarity as a result of training with exemplars was lost in the ensuing discussion and has not properly been exploited. Part of the problem, as Kuhn admitted, was his own carelessness in the use of the paradigm word. It also stems from confusions generated by the difficulty of applying what is primarily an exercise in perceptual learning to the non-perceptual, highly complex, symbolic realm of scientific theories. In Chapter 3 I listed the sorts of problem his account faces. But I also concluded by suggesting how it might be developed. In Chapter 6 I remarked that it fits well with post-empiricist naturalized epistemology. In recent years connectionist models of perception and learning have become very much the fashion in cognitive science but are only just beginning to be applied to the more complex areas of intellectual life such as theory choice. In this respect Kuhn was too far ahead of his time; indeed he was too far ahead of himself, and he was not well placed to exploit and develop his own insights. As Hoyningen-Huene notes, the "natural" standpoint that may exploit the discoveries of neuropsychology is in tension with the "critical" standpoint.[14] The latter, which is sceptical and internalist, predominated in Kuhn's thinking and led him to emphasize its Kantian nature. He regretted the earlier empirical emphasis of his work and sought to reach the same conclusions, in a Kantian manner, from first principles.[15] This was Kuhn's wrong turning. The result is a scepticism that has more in common with Descartes, Berkeley, Hume and Mach, as well as Kant than it has with the post-empiricist epistemologists and philosophers of science working today.[16] Had the natural standpoint defeated the critical standpoint, the empiricist and Cartesian residue in Kuhn's thought might have withered and a far more radical picture would have emerged in its place.

Notes

Preface

1. It could be said that because of the importance of *The Structure of Scientific Revolutions* [hereafter *SSR*] and its impact on the focus and organization of the philosophy of science, the book itself (as a whole, not just its philosophical elements) became a paradigm for philosophy of science and so induced a shift in the very meaning of "philosophy of science" (potentially leading to Kuhnian incommensurability), which allows theoretical history of science to be (or to be very close to) "philosophy of science".
2. For an illustrative natural scientific case see p. 67.
3. See pp. 182–4.
4. Interestingly, the citation of philosophers increases in proportion in the Postscript 1969 to the second edition. And Kuhn's later work is yet more concerned with philosophers and philosophy. As I mention, Kuhn's professional career shifted from history to philosophy. The lack of philosophical references is not itself conclusive evidence of a work not being philosophy – Wittgenstein's writings are notorious in this respect. Nonetheless, I conjecture that Kuhn might have looked back on his earlier work as being more philosophical in conception than it really was, in a way not unlike the misreading of his earlier thinking that Kuhn attributes to Planck (Kuhn 1978).

Chapter 1: Kuhn's context

1. In Ptolemy's system some motion is not uniform about the centre of the circle but only around a mathematical point in space known as the *equant*. See pp. 21–4 for a discussion of Ptolemaic and Copernican cosmology.
2. See Cohen (1985) for a detailed study of the concept of revolution as applied to science, and Toulmin (1970: 41) on continuity in political revolutions, in the context of Kuhn's account of scientific revolutions.
3. Kuhn also adopted an evolutionary analogy for scientific development. See pp. 211–14.
4. Logical empiricism is sometimes taken to be synonymous with logical positivism. Here I shall regard logical empiricism as a slightly broader set of philosophical opinions, roughly those that support the Old Rationalism. It is then fair to regard Popper as a logical empiricist even though he was not a logical

positivist. He did not share the positivist predilection for the philosophy of language or their total rejection of metaphysics. Nor did he like their instrumentalism about theories. A narrower group yet is the Vienna Circle – a loose association of philosophers based in Vienna whose outlook was strongly positivist in the tradition of Mach, and who were influenced by Ludwig Wittgenstein's *Tractatus Logico-Philosophicus* and the logic of Frege and Russell. The Vienna Circle started in 1907 and persisted until the mid-1930s when Nazism led to their emigration, like that of other similarly minded philosophers such as Popper and Reichenbach, to the US and to the UK.

5. What the "phenomena" actually are is a matter over which empiricists do not always agree, for some, such as Otto Neurath they are observable features of the physical world, for others, such as Mach, they are immediately knowable private sensations. See below *n*. 11.

6. See *n*. 4 above.

7. Carnap (1928).

8. Carnap (1963: 78). See Carnap (1939: §§23–5).

9. See Carnap (1939: §§23–5).

10. Fleck (1979).

11. These sense-experiences are typically conceived of as private, inner-states, such as Mach's sensations. As mentioned in *n*. 5, Otto Neurath suggested that the phenomena, and so the subject matter of protocol-sentences, could include features of the subject's immediate environment. But this view was largely rejected by other logical positivists.

12. Or, in Carnap's inductive logic, the statements are used as the basis of the assessment of the logical probability of an inductive conclusion. Since knowledge of the truth of the protocol sentences is in the first place subjective, so is knowledge of the ensuing logical probability.

13. Fleck (1979: 22).

14. *Ibid.* (p. 39).

15. *Ibid.* (p. 41).

16. *Ibid.* (p. 48).

17. Fleck also thinks that "subjective" is an inappropriate term. One reason is that "subjective" suggests an individual property, while the thought-style is a property of the collective.

18. *Ibid.* (p. 90).

19. *Ibid.* (p. 92).

20. *Ibid.* (p. 179).

21. *Ibid.* (p. 179).

22. Fischer (1936).

23. Marx (1859/1913: 11–12).

24. For "earthy" and "watery" substances – the natural motion of "fiery" and "airy" substances is upwards.

25. What Kuhn exactly meant by paradigm is a matter of debate, which we shall examine in Chapter 3. Margaret Masterman maintains that there are 21 senses of the word "paradigm" in *SSR*. Hilary Putnam says that the essential elements of a paradigm are key laws and a striking application of them. According to Bill Newton-Smith the key elements are shared symbolic generalizations (theoretical assumptions etc.), models, values, metaphysical principles, and exemplars and problem situations. Kuhn later used the term "disciplinary matrix" to refer to the constellations of scientific commitments shared by a scientific community and reserved the term "paradigm" for an exemplary puzzle solution, as found in a great revolutionary text.

26. Below we shall consider really how much deviation Kuhn does and should allow for. In a "simple" Kuhnian model there should be no genuine disagreement with the paradigm. But this is historically implausible, since medieval astronomers, especially (but not exclusively) Muslim thinkers, were happy to contemplate what John North calls "non-Ptolemaic schemes" – ones that did away with equants and eccentric deferents (North 1994: 198). But once revisions to paradigms are allowed, it becomes unclear what distinguishes "allowable" revisions from revolutionary ones.

27. I say "incipient" because the crisis really erupted only with Galileo. In particular his use of the telescope as an instrument of astronomical research yielded anomalous observations that it was near impossible for Ptolemaic astronomy and Aristotelian physics to accept – the moons of Jupiter and the phases of Venus above all. Furthermore his careful, well-written and well-publicized reasoning undermined the basis of Aristotelian arguments against the Copernican system. (I have noted above that by contrast Copernicus himself thought that he could argue for his view *within* a broadly Aristotelian physics.) Another example of a science in crisis would be physics at the end of the nineteenth century and beginning of the twentieth. Maxwell's electromagnetism required the existence of an aether, but the Michelson-Morley experiment failed to detect one. (At least it failed to detect an aether that would not drift with moving physical objects.) Rutherford discovered that electrons are not static parts of a solid atom but instead rotate about a nucleus. But then physics could not explain why the moving electrons did not just radiate away their energy and collapse into the nucleus. These seemed to be not simply unsolved puzzles but deeply worrying theoretical problems.

28. *SSR* (p. 153).

29. The existence of incommensurability might seem to make a mystery of the possibility of the continuity on which I remarked at the beginning of this paragraph. This is one of the reasons why Kuhn's emphasis on continuity has been overlooked. I discuss this tension in Chapter 5.

Chapter 2: Normal and revolutionary science

1. As mentioned in *n.* 25 on p. 282, Margaret Masterman (1970: 63) famously claimed to have identified 21 different senses of "paradigm" as used in *SSR*. This claim is however, at least as it is often reported, grossly exaggerated. A variety of different applications, explanations and exemplifications does not show a multiplicity of distinct *senses*. As Masterman concedes, "It is evident that not all these senses of 'paradigm' are inconsistent with one another: some may even be elucidations of others" (Masterman 1970: 65). But as she goes on to say, this multiplicity of uses and explanations raises a legitimate question: Is there anything they have in common that constitutes an explanatorily useful concept?

2. *SSR* (p. 10). Kuhn himself seems to have changed his mind on the question of whether immature science is science without a paradigm. In the first edition of *The Structure of Scientific Revolutions* he talks of "pre-paradigm" periods (e.g. *SSR*: 47, 96, 163), but in the Postscript 1969 in the second edition he states that members of the competing schools in an immature science do operate within paradigms. There he says that the transition to maturity is a matter of permitting the pursuit of puzzle-solving normal science (*SSR*: 179). This is why I have preferred to use the term "immature" rather than "pre-paradigm".

3. *Ibid.* (pp. 11–12).
4. Hoyningen-Huene (1993: 190).
5. *Ibid.* (p. 191).
6. Kuhn (1963: 357). Kuhn here refers to "electricians" – eighteenth-century electrical researchers – where I have generalized the quotation to scientists.
7. Electrical and magnetic phenomena were known to the ancient Greeks. But they became the objects of scientific interest only in the sixteenth and seventeenth centuries.
8. Some periods of immature science may be quite brief, such as that in electrical science before Franklin's paradigm. Others may be rather longer. Some of the human and social sciences may still be in periods of immature science that have lasted for more than a century. The persistence of differing schools of psychoanalysis, springing from the competing work of Freud, Jung, Adler, Fromm, Rank and Klein, suggests that it has not emerged from an immature stage that began over 100 years ago.
9. This statement assumes that it is possible to pinpoint fairly precisely the beginning and end of a revolutionary phase. This in turn assumes that there is a clear distinction between normal and revolutionary science – an assumption we shall question later in this chapter. The revolutions that overthrew classical physics in favour of relativity theory and quantum mechanics lasted a few decades at most, or perhaps just a few years on a more liberal understanding of normal science. It is sometimes said that "The Scientific Revolution", which replaced Aristotelian-Ptolemaic cosmology with Keplerian-Newtonian cosmology, took two centuries. While there is something in this claim, it may also be misleading. The Scientific Revolution might best be seen as a series of related revolutions.
10. Kuhn (1963: 360).
11. *SSR* (p. 30).
12. It may be that to the scientists themselves the distinction between experimental laws and theory extensions may not be so clear.
13. Hoyningen-Huene (1993: 179).
14. This is the case even though Newton's laws were able to explain the relationship after it was known.
15. *SSR* (p. 36).
16. *Ibid.* (p. 53).
17. Hoyningen-Huene (1993: 202).
18. *SSR* (p. 92).
19. *Ibid.* (p. 66).
20. *Ibid.* (p. 93).
21. An alternative proposal might involve a tripartite division of science: normal science; extraordinary science (covering anomaly-driven discovery); revolutionary science.
22. Kuhn quotes Alfonso X as saying that if God had consulted him on the design of the universe, He should have received good advice (*SSR*: 69).
23. *Ibid.* (p. 43).
24. The theory need not be entirely original. Copernicus's heliocentric system had been anticipated by Aristarchus. But what made Copernicus successful where Aristarchus had not been was that Copernicus's ideas coincided with a crisis within the existing view. Similarly, Dalton's atomism had been presaged by Democritus' metaphysics and by the corpuscularianism of the seventeenth century. But in neither of the earlier cases was atomism an answer to an existing, pressing scientific problem. We could characterize an

"idea that is before its time" as an "idea for which no crisis has yet shown the need". Ideas that had delayed receptions due to an absence of a problem for them to solve include Mendel's genetic theory of inheritance and van't Hoff's idea of the asymmetrical carbon atom.

25. *SSR* (p. 84).
26. Cohen (1985: 414).
27. Planck (1949: 33–4).
28. Cohen (1985: 414).
29. Kuhn (1959); *SSR* (p. 151); Kuhn (1963).
30. *SSR* (p. 152).
31. For explanation and discussion, see Chapter 5, pp. 164–5.
32. *SSR* (p. 149). Whether Kuhn's point can be made to stick is a question we shall examine in Chapter 5.
33. Sulloway (1996).
34. *SSR* (p. 82). This is not to deny that the difference may be a psychological rather than a logical or semantic one.
35. *Ibid.* (p. 82).
36. For an extended discussion of the word "revolution" as applied to science see Cohen (1985), especially pp. 51–76.
37. *SSR* (p. 66).
38. *Ibid.* (p. 84).
39. For the basic ideas and applications of catastrophe theory, see Saunders (1980).
40. *SSR* (p. 81).
41. Grant (1852: 61). Also Pannekoek (1961: 304).
42. *SSR* (p. 84).
43. The difference between this concern and that discussed in the preceding section is that the latter asked about the shape of the distribution of instances of scientific change when measured along the normal-revolutionary scale. The concern discussed here asks about the absolute dimensions of the scale. Do the bulk of the instances fall within a fairly narrow band, or are they spread out?
44. The arguments sketched in this paragraph and the next are reinforced in Chapters 3 and 5.
45. Kuhn is quicker than most to see theoretical revision as requiring conceptual revision. For a discussion of his views, see p. 163ff.
46. *SSR* (p. 149). Of course, both advances involved the significant *addition* of new concepts to chemistry.
47. William Harvey's revolution in physiology was not precipitated by a crisis although it was preceded by some important discoveries that were anomalous from the point of view of Galen's teachings. But it is unclear to what extent those teachings formed a coherent paradigm. Cohen denies that there was a Galenic system (Cohen 1985: 189–90). It was thus open to researchers to describe their discoveries as corrections to Galen. Harvey took himself to be instituting a new system of physiology, which included many more discoveries showing Galen to be wrong. But such discoveries did not precede his revolution, bringing it about via a crisis. Rather they were *part* of his revolution. Moving to more recent times, another revolution not precipitated by crisis, according to Rachel Laudan, was experienced by geology in the 1970s (Laudan 1979). Once we allow minor revolutions$_K$ into the picture, many are not preceded by crises, since they come about serendipitously or thanks to the inventions of new technologies. Galileo's discovery of the moons of Jupiter

and, yet again, Roentgen's X-rays are like this. These may have in turn *caused* crises but were not themselves *caused by* any crisis.
48. *SSR* (p. 96).
49. Kuhn asserts his belief in completeness from the outset, "The commitments that govern normal science specify not only what sorts of entities the universe does contain, but also, by implication, those it does not" (*Ibid.*: 7). He reiterates it elsewhere: "For the historian, or at least for this one, theories are in certain respects holistic. So far as he can tell, they have always existed (though not always in forms one would comfortably describe as scientific), and they then always cover the entire range of conceivable natural phenomena (though often without much precision)" (Kuhn 1977b: 20).
50. Interestingly Kuhn nowhere mentions the discovery of the structure of DNA, although Cohen points out that it is second only to advances in computing in evoking the name "revolution" in the press (Cohen 1985: 384–5). Mention of the computer suggests another case of a non-revisionary revolution, which although partly technological was also genuinely scientific. Mathematics itself is a field that experiences revolutions almost all of which are conservative, although there are notable exceptions.
51. Kuhn acknowledges the existence of mini-revolutions, for example in *SSR* (p. 49), but then adds, "it is still far from clear how they can exist". See also *Ibid.* (pp. 180–81). He later discusses proliferation but does not discuss in any detail how it fits in with the theory of paradigms, normal science and revolutions (1992: 17–20).
52. This is not to say that we can only spot revolutions with the benefit of hindsight, although only hindsight will be conclusive. It may be obvious at the time of its occurrence that a revolution such as Einstein's would have a huge influence on subsequent science.
53. A point Kuhn clearly recognizes – but by seeing *crisis* in relative terms (*SSR*: 92–3).

Chapter 3: Paradigms

1. This example is taken from Bloor (2000b).
2. It is possible for a scientist to be correct in identifying the regularity but mistaken in the explanatory theory. Hence the two merit independent discussion, which is why in this chapter on Kuhn's theoretical explanation of scientific change, I shall for the most part assume that he was right about the pattern of that change, my criticism at the end of the last chapter notwithstanding.
3. Hoyningen-Huene (1993: 133–4).
4. Kuhn (1974: 460; 1977a: 294).
5. *SSR* (p. 175).
6. Kuhn (1974: 463); *SSR* (pp. 181–7).
7. *Ibid.* (p. 187).
8. *Ibid.* (p. 43).
9. Hoyningen-Huene uses the phrase "lexicon of empirical concepts".
10. On the lack of a clear discovery/justification distinction see Bird (1998: 260–62).
11. Cf. Lakatos: "For Kuhn scientific change . . . is a mystical conversion which is not and cannot be governed by rules of reason and which falls totally within the realm of the *(social) psychology of discovery*" (Lakatos & Musgrave 1970: 93).
12. *SSR* (p. 191). Cf. quotation from Lakatos in *n.* 11 above.
13. *Ibid.* (p. 192).

14. *Ibid.* (pp. 192–3).
15. *Ibid.* (p. 180).
16. Musgrave (1971: 290).
17. See Kuhn (1978).
18. In an extended discussion of the relation between the components of a disciplinary matrix, Paul Hoyningen-Huene comes to a slightly different conclusion. He sees the various components as *"linked moment of single unity"* (Hoyningen-Huene 1993: 157). The core of his argument as regards symbolic generalizations and metaphysical and heuristic models is that the content of these generalizations and models is incomplete without the exemplars. It is certainly true that Kuhn treats symbolic generalizations as empty formalisms that say something about the world only in virtue of concrete applications in exemplars. Because of this unity Hoyningen-Huene does not think that we have to see any component as the central "moment", regarding that as a "rather vague question". By contrast I do think it is important to identify exemplars as having an explanatorily central role, particularly because failure to do so may rob, as discussed above, the notion of paradigm of its explanatory power. However, I do concur with Hoyningen-Huene's view that it is the linkages between the components that are responsible for the breadth of usage in *The Structure of Scientific Revolutions*, while I can add that identifying exemplars as explanatorily central also explains his emphasis on them in 1969 and thereafter in a way that makes later use of the disciplinary matrix concept redundant – Kuhn doesn't use it after 1969.
19. An unusual puzzle solution might work by "dissolving" as opposed to solving the anomalous puzzle, as relativity did for the puzzle of the undetected aether. In this way new puzzles will differ from the old ones.
20. Symbolic generalizations are like watered-down theories – they have mathematical or logical form but no further content. This is because, Kuhn thinks, the exemplar provides the symbols with meaning. Since we are separating out the semantic function from exemplars we need not worry about this difference for current purposes.
21. *SSR* (pp. 188–9).
22. *Ibid.* (p. 189).
23. *Ibid.* (p. 169).
24. Conceivably our ability to recognize and use some such patterns is innate.
25. A pattern of reasoning, if it can be articulated, can always be *expressed* as an assumption; but that does not mean that it is best conceived of as an unstated assumption. And it is not obvious that every pattern of reasoning can be articulated.
26. Kuhn (1970b: 234).
27. In the next chapter we shall look at a study carried out by Jerome Bruner and Leo Postman that illuminates the perceptual aspect of habituation. Of this Kuhn wrote: "Either as a metaphor or because it reflects the nature of the mind, that psychological experiment provides a wonderfully simple and cogent schema for the process of scientific discovery" (*SSR*: 64). It makes a considerable difference to Kuhn's explanatory use of the exemplar notion which of these is the case.

Chapter 4: Perception and world change

1. Logical empiricists rarely laid out the structure of their views in anything like this way. Not all endorsed all these assumptions – as already mentioned, Neurath did not, but was therefore regarded as backsliding by his colleagues.

Popper, not being a positivist, was like Kuhn in rejecting (iv). Popper did however accept at least (i). Kuhn, I shall be arguing, continued to hold (i), (ii), and (iii).

2. The independence thesis is related to what Feyerabend calls the "stability thesis". Feyerabend attacks this thesis as being central to the positivist picture, but Preston argues that not all positivists need adopt it (Preston 1997: 33–4).
3. Hanson (1965).
4. See p. 14.
5. Although I use the terminology "strong" and "weak", this does not denote a precise logical relationship, since "strongly seeing X" does not entail "weakly seeing X". This is because, for example, objectual seeing is non-intensional, and so one may strongly see X without weakly seeing X, if one does not have the concept "X". Nonetheless, I use the terminology because the weak senses are neutral as regards truth and existence while the strong senses are not.
6. Note that I am not claiming that one can strongly see despite having *no* relevant concepts at all. To establish that would take some further discussion. Rather I am saying simply that it can be correct to say that S sees X even though S does not have *that* concept "X".
7. Or, were Tycho right after all, that the Earth is stationary, then there is no moving Earth – nonetheless, Kepler weakly sees a moving Earth.
8. The difference between "intensional" and "extensional" contexts is illustrated by a logical difference between the verbs "to believe" and "to kill". The *intension* of a word is its sense or verbal meaning; its *extension* is the set of things in the world the word denotes or describes. And so, very roughly, the truth values of sentences involving extensional contexts are sensitive to differences only in the things denoted, while intensional contexts create a sensitivity to verbal meaning as well. Hamlet ran his sword through the arras (a kind of curtain) killing a man he thought was his uncle the king. But in fact the victim was Polonius the Lord Chamberlain. The sentence "Hamlet killed the man behind the arras" is therefore true. And since the man behind the arras is Polonius, it follows that the sentence "Hamlet killed Polonius" is also true. However, a parallel move is not valid when the context involves believing rather than killing. The sentence "Hamlet believed that the man behind the arras is the king" is true, but this time we cannot swap "Polonius" for "the man behind the arras", since to do so gives us the sentence "Hamlet believed that Polonius is the king", which is false. The word "killed" creates an extensional context: the truth value (true or false) of the sentence "Hamlet killed x" depends just on the thing in the world that "x" denotes. But "believed" creates an intensional context: the truth value of the sentence "Hamlet believed x is the king" depends not on the thing "x" denotes but instead on the verbal meaning of what is put in place of "x". Since "the man behind the arras" and "Polonius" do not have the same meaning Hamlet need not know that they have the same denotation, and so we may not get the same truth value when we put "Polonius" in place of "x" in "Hamlet believed x is the king" as when we substitute "the man behind the arras". Related to the sensitivity of intensional contexts to changes in meaning is the fact that they are also sensitive to the availability of relevant concepts to the subject. Thus if Hamlet does not have the concept "acupuncturist" he cannot believe that Polonius is an acupuncturist; nonetheless, Hamlet can kill an acupuncturist without having that concept. (Note that "intensional" has an "s" and is different from "intentional". But the meaning of "intensional" may be remembered

by reflecting that "intends" creates an intensional context: Hamlet intended to kill the man behind the arras but did not intend to kill Polonius.) The difference between intension and extension arises again in Chapter 5.

9. Hanson (1965: 20).
10. See *n*. 12 below for objections to this common view.
11. This is not to say that Hanson and the traditional empiricist agree on how to characterize the internal component. For the latter it is something like: the experience which is as of (strongly) seeing – what is common to seeing, hallucinating, imagining and so on. See p. 103.
12. Experiences can be very similar to one another. Let A and B be incompatible experiences that are sufficiently similar that a subject S cannot discriminate having A from having B, for instance the experiences of two very similar shades of colour. We cannot then rule out the possibility that S might have experience A but believe she was having B. In such circumstances she would have a false belief about her experiences and so not knowledge. Even if she got it right, and believed she was experiencing A, she would still not have knowledge of that, for she cannot discriminate being in a situation of experiencing A from being in a situation of experiencing B. And if she cannot make this discrimination, she cannot know she is in the one situation rather than the other. See Williamson (1996).
13. It is true that weak perception can be true, perhaps by chance. In fact we should want our inferences to start not merely from what is true but rather from what is known. In which case we must have strong seeing rather than merely accidentally true weak seeing, since strong seeing entails knowing. See Williamson (1997).
14. *SSR* (p. 113). Hanson is the philosopher most frequently cited in *The Structure of Scientific Revolutions*.
15. *Ibid.* (p. 111ff.).
16. *Ibid.* (p. 118).
17. *Ibid.* (p. 132).
18. *Ibid.* (pp. 118–19). Although Kuhn does not here *say* that the Aristotelian *saw* a stone falling with difficulty, it is clear from the subsequent discussion and from other examples that he was not avoiding this locution.
19. Bruner & Postman (1949).
20. *SSR* (p. 112).
21. *Ibid.* (p. 114).
22. Hanson (1965: 19).
23. The issue under discussion here, and in this chapter more generally, is related to the Sapir-Whorf hypothesis in linguistics and psychology. That hypothesis states that a person's language is a major factor in determining their world-view, thoughts and perceptual experiences. Psychological tests show that strong versions of the hypothesis are false, but that weak versions have some truth in them. There is of course an obstacle in the way of testing the hypothesis, since differences in language often reflect differences in environment, in expected use, and hence in the experience of a typical user of the language. And therefore it is difficult to tell whether the psychological and perceptual differences between users of different languages reflect the differences in their languages or differences in their experiences of the world (see Lyons 1981: 303–12). Kuhn mentions Benjamin Whorf in the Preface to *The Structure of Scientific Revolutions* (*SSR*: vi).
24. Similarly it is clear that Hanson's use of the Tycho–Kepler case is not metaphorical, since it is part of his discussion of observation.

25. *Ibid.* (pp. 118–19).
26. *Ibid.* (p. 120).
27. It may be that there are other significant psychological differences between Tycho and Kepler that although not perceptual are nonetheless not merely a matter of differing belief. I make some suggestions below. For similarity relations see p. 74.
28. *Ibid.* (p. 132).
29. *Ibid.* (p. 116).
30. Not every ostensive definition brings a change in experience. For we might already have had the relevant discriminatory capacity, which the ostensive definition exploits in giving us a new concept. Since the discriminatory capacity was already present, there is no change in perceptual experience.
31. See Broad & Wade (1985) for accounts of Blondlot's N-rays and Rosenthal's study.
32. And in any case the data were extracted from photographic plates, so the relevant perceptions would be perceptions of the plates. However, it should be added that there is every reason to think that the process of obtaining the data from the plates was far from unequivocal – Eddington's preferences may well have influenced this process (see Collins & Pinch 1993: 54–5). This however has nothing to do with the current discussion of *perception*. It may well be pertinent to *observation*, for scientists would naturally mean the processed data by the phrase "Eddington's observations" and not the photographic plates nor his perceptions of them. As Collins and Pinch put it "the Eddington observations were not just a matter of looking through a telescope and seeing a displacement; they rested on a complex foundation of assumptions, calculations, and interpolations from two sets of photographs" (*Ibid.*: 47). Accepting this gap between perception and observation is also to reject the empiricist view of observation, but by rejecting assumption (ii), the experiential basis, rather than the independence assumption. While this may well be Kuhnian in spirit it does not enter into Kuhn's discussion of observation, which, I suggest, is still influenced by an empiricist conception of observation as perceptual.
33. *SSR* (p. 150).
34. See, for example, *Ibid.* (p. 118) for the connection between perception and world change.
35. *Ibid.* (p. 121).
36. Kuhn's conflation of perception with sensation is itself an empiricist or more generally a Cartesian characteristic. See Hacker (1987) for an extended discussion.
37. Locke also includes a third class of quality, tertiary qualities, that are powers of things to produce changes in other things.
38. Hoyningen-Huene (1993: 35).
39. *Ibid.* (p. 33).
40. *Ibid.* (p. 34). He might have added that adherents of different paradigms by and large have similar experiences.
41. *SSR* (p. 121).
42. See for example Mach's account of sensations on p. 10.
43. Kuhn (1979a: 418–19). Hoyningen-Huene takes this silence to be evidence that Kuhn puts them on the object side. But equally it could be taken to indicate that having earlier been happy to put them on the object side, he is no longer so sure.
44. Kuhn (1991a: 12). It should be mentioned that the later papers where the Kantianism is most explicit were written after his discussions with

Hoyningen-Huene and the publication of the latter's book where the Kantian parallels are carefully described.
45. Kuhn (1979a: 1992).
46. *SSR* (p. 195). See p. 225.
47. Hoyningen-Huene (1993: 33); *SSR* (p. 118).
48. Kuhn (1974: 473–82). See pp. 194–5.
49. The connections studied by connectionism are neural, the connections and associations being discussed in this section are psychological. The former support and explain the latter.
50. The term "rational" must here be understood in a robust way, not just as a rhetorical device.
51. *SSR* (p. 192).
52. *Ibid.* (pp. 192–3). Visual exemplars, that is. Kuhn seems to think that non-visual exemplars can have the same role, but I have argued that they do not.
53. *Ibid.* (p. 192).
54. *Ibid.* (p. 192).
55. Hoyningen-Huene (1993: 45–6).
56. Even if the solution is a revolutionary one.
57. *SSR* (p. 192).
58. Van Fraassen (1980).
59. *SSR* (p. 114). For van Fraassen not every instrument is acceptable as a means of visual observation, only those that enable us to see things that could in other circumstances have been seen with the naked eye. Hence telescopes enable observation but microscopes do not.
60. I shall be talking primarily about choice of theories within a paradigm, except where stated otherwise.
61. Van Fraassen (1980: 112). Conversely Kuhn briefly alludes to their affinities (1991a: 6–7).
62. Although in the preface he acknowledges that there is more to be said about them (*SSR*: ix–x).
63. Elsewhere I have called these "meaning" and "cognitive" dependence of theory on observation (Bird 1998: 173–5).
64. Van Fraassen (1980: 14).
65. This is somewhat controversial.
66. *Ibid.* (p. 19).
67. Kuhn does say things that lead in this direction, for example: "The operations and measurements that a scientist undertakes in the laboratory are not 'the given' of experience but rather 'the collected with difficulty'" (*SSR*: 126). What needs recognition is that what normally is called "observation" in science is closer to the latter than to the former.
68. See Williamson (1997) for the view that one's evidence is precisely what one knows. On this view one's evidence is not essentially observational unless one's knowledge is too. If theoretical knowledge is possible, then so is theoretical evidence.

Chapter 5: Incommensurability and meaning

1. The incommensurability thesis was introduced by Feyerabend at almost the same time as Kuhn (Feyerabend 1962). While Feyerabend's view of incommensurability remained unchanged, Kuhn's underwent some development.
2. Newton-Smith, for example, takes the incommensurability thesis to be a thesis about non-comparability: "The thought that theories are incommensu-

rable is the thought that theories simply cannot be compared and consequently there cannot be any rationally justifiable reason for thinking that
one theory is better than another" (1981: 148). Shapere says something
similar: "if two theories (paradigms) are incommensurable, they cannot be
compared directly with one another" (Shapere 1966, in Hacking 1981: 55).

3. Say that this were false, and we found that for some unit the sides were a
units long and the hypotenuse was b units long, where a and b are whole
numbers. Then it would follow that $\sqrt{2}$ is equal to the rational number b/a,
which would contradict Euclid's proof that $\sqrt{2}$ is irrational.

4. Kuhn (1983a: 670).

5. The positivists in particular would want these statements to be reports of
experiences of looking at the pointer.

6. Kuhn (1970b: 266).

7. *Ibid.* (p. 266).

8. There are logical problems with this idea, but we shall pass over these. See
Miller (1974), Newton-Smith (1981), Bird (1998) for details.

9. Preston (in a discussion of Feyerabend) says otherwise: "If I understand
language A, then I cannot be credited with an understanding of another
language B, unless I can generally perform translations from B into A and
vice-versa" (1997: 188). Certainly being able to perform translations is
evidence of understanding, but it is not a necessary condition. An ability to
translate may be expected only if translations are possible. And we could
imagine a person able to communicate fully in both of two languages but who
has difficulty using both at the same time, as if there were room for only one
language at a time in his active mental module for language. Such a person
would understand two languages between which he could not easily translate. This is fairly common among bilingual children.

10. Kuhn (1970b: 268).

11. According to Quine the existence of a thing requires determinacy of identity,
and the indeterminacy of translation thesis supposedly shows that there is no
determinacy of identity for meaning. Hence there is no meaning.

12. *Ibid.* (p. 268).

13. *Ibid.* (p. 269).

14. It might be reasonable to translate "Ishtar" as "Venus" if in some appropriate
sense the Romans and the Babylonians were worshipping the same goddess.
However it is not clear that "Venus" *now* is the name of a goddess, and
certainly the noun phrase "the planet Venus" is not.

15. The origin of the world "planet" is the Greek πλανητης meaning "wanderer".

16. Strictly speaking the extension may change with no change in intension if
new planets come into being (as opposed to being discovered). We may either
safely ignore this complication, or deal with it by thinking of intensions and
extensions in an tenseless way.

17. This positivist view of theoretical meaning is discussed on pp. 10–13.

18. It would not be *necessary* that the extension has changed, since two different
intensions can share the same extension.

19. Kuhn gives the impression that Berthollet's was the standard view. But
according to Brock the reverse is closer to the truth: the implicit assumption
of many chemists was that for a chemical reaction the proportions of the
reactants is constant. See Brock (1992: 144).

20. *SSR* (pp. 132–3).

21. *Ibid.* (p. 133).

22. At least if the range of application of Newton's laws is limited. It would

remain the case that certain important additions to Newtonian mechanics, such as the theory of the aether, would need revision.
23. *Ibid.* (p. 102).
24. Kuhn (1990).
25. *Ibid.* (pp. 307–8).
26. *SSR* (p. 102).
27. I am here following David Lewis's terminology of sparse and abundant universals (Lewis 1983: 346–6; 1986: 59–63).
28. The Queen has a sister but no brothers.
29. For "mass" there are further complications too. Mass, along with concepts like charge, temperature, momentum and so on is the concept of a variable quantity, what I shall call a *quantity-property*. With the other properties already discussed (being a compound, being deciduous, being gold – let us call them quality-properties) we typically ask "Does such-and-such have this property?" but with the quantity-properties we ask instead "What mass does such-and-such have?", "How much charge does it have?", "How great is its momentum?" The question regarding the quality-property is whether the object is within the extension of the term. But with the quantity-properties, the question seeks the value of the property possessed by the object.
30. That there is supposedly successful but distinct reference is shown by the passage quoted on p. 165.
31. Kuhn (1990: 306).
32. It would still have to be the only entity satisfying that proportion.
33. Kuhn (1983c: 566–7).
34. I describe a thin version of intensionalism – causal descriptivism – in greater detail on p. 185ff.
35. The fact that they do not translate each other does not preclude their both being translatable using some common set of terms. This will be discussed below.
36. It is sometimes thought that the existence of subatomic particles shows that the definition of "element" in terms of non-decomposability is erroneous. This is not quite right, since such particles are not substances in the relevant sense. The fact that elements may be separated by mechanical means into their isotopes is more of a threat to this definition.
37. Putnam (1975b); Kripke (1980).
38. Kuhn (1990: 310).
39. *Ibid.* (p. 312).
40. Sankey (1997b).
41. Lewis (1970).
42. It is true that as a result of a theory change we may realize that a term has no reference and thus no use for us, and so we may recycle the term with a completely new meaning, or the term may revert to a general meaning and lose its special scientific significance. This again may cause confusion. So when today we read the title of Copernicus's *On the Revolutions of the Spheres* we naturally take "spheres" to refer to the planets, whereas Copernicus, being an Aristotelian and a reformer of Ptolemy was in fact trying to refer to what he believed to be the crystalline globes whose motion carries the planets. But such difficulties of ambiguity do not follow from every theory change, and in any case are a rather more general feature of language.
43. Sankey (1997b).
44. or the generalization of that, $\exists! x_1, \ldots x_n \, T(x_1, \ldots x_n)$, where $x_1, \ldots x_n$ are variables taking the places of all the n theoretical terms in T.
45. Papineau concedes that the intension of a theoretical term is not precise,

since it may not be determinate quite what and how much is included in T_{core}. He argues that this should not bother us. One reason for the indeterminacy is precisely that there is the room for manoeuvre without upsetting the unique fixing of reference. Correspondingly, from the point of view of the scientist who is more interested in reference than intension, we typically just do not need to worry about such indeterminacy of meaning, even in cases of significant theory revision. And the reason the scientist is interested in reference is that this is what is required for theory comparison. If the translator finds that translation is indeterminate, that is just a local difficulty for them, not for the rest of us. Indeterminacy (or imprecision, as Papineau prefers to say) in meaning need not lead to indeterminacy of reference. This concession is not the same as loose intensionalism, since descriptions that clearly fall within the intension must still be satisfied if the term refers.

46. Kuhn (1983a: 682–3).
47. Sankey (1998).
48. Kuhn (1987: 19–20).
49. Kuhn (1974: 473–82).
50. Actual extension need not be the same as the apparent or even intended extension. Sometimes Kuhn writes as if the extension of a term is just what the theoretician believes it to be.
51. It is a restriction that again emphasizes how much Kuhn inherits from empiricism.
52. As was done when a new taxonomy of monetary value was introduced to the UK, along with decimal currency, where ambiguity in "pence" was avoided by talk of "old pence" and "new pence".
53. Note that I am *not* saying that we could combine the French and English languages; the different syntaxes of the languages prevent that. But we can – and do – import words and their associated taxonomies from other languages. Creoles are often formed from the vocabularies of different languages in this way.
54. Kuhn (1991a: 4).
55. *Ibid.* (pp. 4–5).
56. Sankey (1998).
57. Hacking (1993). For a predecessor to Hacking's account of taxonomy, see Thomason (1969).
58. That is, if **a** stands in K to **b**, and **b** stands in K to **c**, then **a** stands in K to **c**, and if **a** stands in K to **b**, then **b** does not stand in K to **a**.
59. See Bird (1998: 96–8, 104–11) for discussion of such examples. David Wiggins also argues that cross-classification is inevitable in areas as diverse as Chomskyean linguistics and biology (Wiggins 1980: 201–3).
60. Kuhn (1993: 318).
61. For this reason it looks as if we have not two taxonomies but just one.
62. I did mention another case where a shift of non-empty extensions is possible, where the theories that define the intensions are both true, but different, before and after the shift. Since the theories are both true the intensions themselves cannot licence conflicting expectations. According to some views of the meaning of "water", such as given by Mellor (1977), something like this did happen.
63. Kuhn (1991a: 5). Kuhn takes this example from Lyons (1977: 237–8).
64. Kuhn remarks that for him the origin of the notion of incommensurability was a matter of sensitivity to the meanings of historical texts (Kuhn 1991a: 4). The writings of scientists of earlier generations may seem confused, irrational, or just obviously wrong. A recognition that terms may have

changed their meaning, that the most obvious translation may not be the correct one, is needed for historical accuracy. A better knowledge of their theories, their background beliefs and cultural context may enable an understanding of the relevant passages that makes them seem less confused, more rational and not so obviously mistaken. This is indeed important and is perhaps especially salient for historians of science, the majority of whom, like Kuhn himself, have had a scientific rather than historical education. The lesson is perhaps less urgent for philosophers, both practically and theoretically. Historians of philosophy, especially ancient philosophy, often bend over backwards to interpret their authors in the most favourable light; Davidson's principle of charity provides a theoretical exhortation of just this kind. (Even charity can be taken too far. As we have seen, Kuhn's very thick intensionalism threatens to require us to interpret every theory as true.)

Chapter 6: Progress and relativism

1. *SSR* (p. 206).
2. *Ibid.* (p. 169).
3. This is a rather simplified picture of what is called an "arms race" in evolutionary biology. There are of course many other factors determining evolution, not least being various fixed physical and physiological limits.
4. Those who are sceptical about the extent of possible evidence will think that a theory is unfalsifiable when it is empirically adequate. Van Fraassen is such a sceptic, and so too, I have argued, is Kuhn. But in that case empirical adequacy is the optimal state towards which theories may evolve.
5. Barnes & Bloor (1982).
6. Newton-Smith (1981: 237–65).
7. Sulloway (1996). For a philosophical discussion of the extent and limits of reason see Kornblith (1999).
8. Kuhn (1983c: 563–4; 1992: 11). In the 1983 paper Kuhn then replaces this view with a conception of irrationality as violating norms associated with the concepts "science", "accuracy" and so on.
9. *SSR* (p. 207).
10. Kuhn (1993: 330–31).
11. *SSR* (pp. 44–5); Kuhn (1977a: 121). Kuhn may well have absorbed much Wittgenstein indirectly from discussions with Stanley Cavell at Berkeley (see *SSR*: xi).
12. Wittgenstein (1953: §66).
13. Barnes (1982); Bloor (1976).
14. Wittgenstein (1953: §201).
15. Kuhn (1974: 472).
16. See p. 169 for "grue".
17. Barnes (1982: 31).
18. The disquotational schema does not hold when the "*p*" within the quotation marks on the left is in a different language from the "*p*" outside the quotation marks on the right. Furthermore there are difficulties when the quoted "*p*" contains indexical expressions or ambiguous ones. We can bypass such difficulties by regarding the disquotational schema as conditional as follows: if "*p*" states that *p* then "*p*" is true if and only if *p*. In the examples that follow we can take the truth of the antecedent ("*p*" states that *p*) as knowable *a priori* for English speakers, by taking both the object language and the metalan-

guage to be English and by ensuring that we deal only with unambiguous eternal sentences.

19. There is nothing metaphysically question-begging in my use of "property" here, since it is "abundant" properties I have in mind, which obey their own disquotational and equivalence schemata: "x is C" is true if and only x has the property C; x is C if and only x has the property C. This very weak notion of property does not raise deep metaphysical questions of the kind associated with sparse, "scientific" properties.

20. See Baker & Hacker (1984).

21. Hoyningen-Huene says: "It is not the scientist, conceived of as an individual, who has similarity relations at his disposal; they are rather the property or attribute of particular communities, and the individual has access to them only *qua* member of the community" (1993: 82). We need to be careful how we understand this. It is the individuals that possess, *qua* isolated individuals, the ability to perceive the relevant similarities and dissimilarities. The role of the community is to give the subject the exemplars, to train him or her in their use, and to make sure all individuals have the same training with the same exemplars. In short the community causes the individual to have his or her similarity relations, and causes all individuals to have the same similarity relations. This is consistent with the fact that the criteria for the application of a similarity *predicate* may be set at the community level; and hence is also consistent with the fact that the individual has access to the corresponding similarity *concept* only *qua* member of the community (thanks to the division of linguistic labour).

22. *SSR* (p. 45; original emphasis). Kuhn concedes that Wittgenstein himself says almost nothing about what the world must be like for family resemblance concepts to be possible. Which shows that Kuhn is even further from being a finitist than Wittgenstein.

23. *Ibid.* (p. 195).

24. *Ibid.* (pp. 206–7).

25. Laudan (1981).

26. Kuhn (1992: 14).

27. Even so, I suspect that a realist would be on stronger ground if she construes progress as the increase of knowledge rather than of verisimilitude. For then Kuhn's worry concerning opinions on space-time disappears since we could reasonably argue that the relevant beliefs of Aristotle and Newton (and possibly Einstein too), whether true or not, did not count as knowledge.

28. Hoyningen-Huene (1993: 263–4).

29. Thought about false propositions is also thought about the world. If I think "a gold atom has 78 protons" that thought is still about gold.

30. Kant (1933: 219–20); James (1904).

31. Hempel (1935).

32. Note that this change to the conception of *truth* follows from an *epistemological* problem. It seems to trade on an ambiguity in "a statement is adopted as true if sufficiently supported by protocol statements" which can be understood as reflecting on the standards of acceptance or on the standards of truth.

33. *Ibid.* (pp. 57–8).

34. Kuhn (1991a: 6–8).

35. The equivalence schema is not the same as the disquotational schema, since the former involves no semantic ascent. But the disquotational schema follows from the equivalence schema plus: "p" is true if and only if it is true

that *p*, which like the disquotational schema is trivial *a priori* knowledge for an English speaker when "*p*" is in English.

36. For a full account of minimalism see Horwich (1998). Horwich argues, as I do here, that minimalism is compatible with realism. He also argues that the other main theories of truth are compatible with realism too while I suggest that not all are. See Horwich (1998: 52–67).
37. Kuhn (1993: 330).
38. See pp. 151–2, and Kuhn (1970b: 266).
39. For a detailed discussion of Inference to the Best Explanation see Lipton (1991). For a briefer discussion see Bird (1998: 85–91, 144–60).
40. For an extended discussion of why Inference to the Best Explanation might even generate a normal science-revolutionary science distinction, see Bird (1999).
41. Kuhn's epistemological relativism is expounded and defended in Doppelt (1978).
42. *SSR* (pp. 109–10).
43. Copernicus *De Revolutionibus* quoted in Koestler (1968: 199).
44. *SSR* (pp. 199–200).
45. For further evidence of a non-factive, relative notion of knowledge, see p. 271. Interestingly, Popper is a rarity among those of the Old Rationalist camp in regarding knowledge as non-factive.
46. For an exposition of naturalism in philosophy see Papineau (1993).
47. Descartes was not a sceptic. But this is only because he has an *a priori* argument demonstrating the existence of a benevolent God. God, being good, guarantees the truthfulness of our clear and distinct perceptions. If one removes these elements from Descartes, the remaining position would be sceptical.
48. This threatens a regress – which is why internalism is conducive to scepticism
49. For an argument whose conclusion rejects the K-K principle, see Williamson (1992).
50. This is not to suggest that belief-forming capacities came before knowledge-producing capacities. Rather I suspect that the capacities developed together, the one being a more general version of the other.
51. For more on externalism in philosophy of science and pluralism about method, see Bird (1998: 215–87).
52. Kuhn (1993: 330–31).
53. Kuhn (1991a: 9).
54. Sankey associates this relativism about truth to his thesis that Kuhn is an ontological relativist. See Sankey (1997a: 42–61).
55. Kuhn (1992: 14).
56. Note that this benefit does not accrue to R-truth.
57. This statement can be made even if we accept Kuhn's views about the semantic incommensurability of the theories, for he accepts that relevant measurements may be made in the same ways, for slow speeds (*SSR*: 102).
58. Paul Churchland notes, in passing, the affinity between Kuhn's thinking and naturalized epistemology (Churchland 1979: 123–4).
59. Laudan (1984, 1989); Worrall (1988, 1989).
60. *Ibid.* (p. 274).
61. Kuhn (1992: 11).
62. E.g. Kuhn (1991a: 6), and (1992: 16). Kuhn often talks more circumspectly about "knowledge claims" and sometimes about belief.

63. Kuhn (1991a: 7; 1992)
64. He writes: "an evolutionary epistemology need not be a naturalized one" (Kuhn 1991a: 6).

Chapter 7: Kuhn's legacy

1. Kuhn (1991b: 23).
2. *Ibid.* (p. 17).
3. *Ibid.* (p. 23).
4. Kuhn does mention another reason for doubting that all the social sciences can achieve a state of normal science. When what is being studied is a social or political system there may not be sufficient stability to support a tradition of puzzle-solving research that is inherited from the scientists' immediate predecessors (*Ibid.*: 23–4).
5. Kuhn (1992, 1991a).
6. Bloor (1997: 124).
7. Kuhn (1992: 14, 20).
8. Bloor (1997: 125).
9. *Ibid.* (1997).
10. Bloor (2000a: 3); Mannheim (1953).
11. Fuller (2000).
12. Fuller locates this line of thinking in a tradition going back to Whewell. He cites Planck as a key exponent of it *contra* Mach, and links it (not entirely perspicuously) to the realism-instrumentalism debate.
13. Note that *The Structure of Scientific Revolutions* was published as one book in the series *International Encyclopedia of Unified Science*, whose editor was the positivist Otto Neurath and among whose other editors and advisors were such famous positivists and empiricists as Rudolph Carnap, Philip Frank, Herbert Feigl, Ernest Nagel, Bertrand Russell, Hans Reichenbach and Joseph Woodger. Note also that Kuhn's doctoral advisor was the philosophically minded physicist P. W. Bridgman. Bridgman was famous for promoting "operationalism", one version of positivistic empiricism.
14. See pp. 140–41.
15. Kuhn (1991a: 6; 1992: 10).
16. I have in mind, for example, Dretske, Goldman, Armstrong, Mellor, Papineau, Williamson. Not that all of these have eliminated every last vestige of empiricism.

Bibliography

Works by Kuhn

1957. *The Copernican Revolution: Planetary Astronomy in the Development of Western Thought*. Cambridge MA: Harvard University Press.

1959. The Essential Tension: Tradition and Innovation in Scientific Research. See Taylor (1959), 162–74; and Kuhn (1977), 225–39.

1962. *The Structure of Scientific Revolutions*. Chicago: University of Chicago Press [2nd edn 1970; 3rd edn 1996].

1963. The Function of Dogma in Scientific Research. See Crombie (1963), 347–69.

1970a. Logic of Discovery or Psychology of Research? See Lakatos & Musgrave (1970), 1–20; and Kuhn (1977), 266–92.

1970b. Reflections on my Critics. See Lakatos & Musgrave (1970), 231–78.

1970c. Postscript 1969. See Kuhn (1962 [2nd edn 1970]), 174–210.

1974. Second Thoughts on Paradigms. See Suppe (1974), 459–82; and Kuhn (1977), 293–319.

1977a. *The Essential Tension. Selected Studies in Scientific Tradition and Change*. Chicago: University of Chicago Press.

1977b. The Relations between the History and the Philosophy of Science. See Kuhn (1977a), 3–20.

1977c. Objectivity, Value Judgment, and Theory Choice. See Kuhn (1977a), 320–39.

1978. *Black-body Theory and the Quantum Discontinuity 1894–1912*. Oxford: Clarendon Press [1987 edition with new Afterword, Chicago: University of Chicago Press].

1979a. Metaphor in Science. See Ortony (1979), 409–19.

1979b. Foreword. See Fleck (1935 [1979]).

1980. The Halt and the Blind: Philosophy and History of Science (review of C. Howson 1976 *Method and Appraisal in the Physical Sciences* Cambridge: Cambridge University Press) *British Journal for the Philosophy of Science* **31**, 181–92.

1983a. Commensurability, Comparability, Communicability. See Asquith & Nickles (1983), 669–88.

1983b. Response to Commentaries. See Asquith & Nickles (1983), 712–16.

1983c. Rationality and Theory Choice. *Journal of Philosophy* **80**, 563–70.

1987. What are Scientific Revolutions? See Krüger et al. (1987), 7–22.

1990. Dubbing and Redubbing: The Vulnerability of Rigid Designation. See Savage (1990), 298–318.

1991a. The Road Since Structure. See Fine et al. (1991), 2–13.

1991b. The Natural and the Human Sciences. See Hiley et al. (1991), 17–24.

1992. The Trouble with the Historical Philosophy of Science. *Robert and Maurine Rothschild Distinguished Lecture 19 November 1991 An Occasional Publication of the Department of the History of Science*. Cambridge MA: Harvard University Press.

1993. Afterwords. See Horwich (1993), 311–41.

For a complete bibliography of Kuhn's writings, see Hoyningen-Huene (1993), 273–8, 302.

References

Achinstein, P. 1964. On the Meaning of Scientific Terms. *Journal of Philosophy* **61**, 497–509.

Asquith, P. & I. Hacking (eds) 1979. *PSA 1978. Proceedings of the 1978 Biennial Meeting of the Philosophy of Science Association vol. 2*. East Lansing: Philosophy of Science Association.

Asquith, P. & P. Kitcher (eds) 1985. *PSA 1984. Proceedings of the 1984 Biennial Meeting of the Philosophy of Science Association vol. 2* East Lansing: Philosophy of Science Association.

Asquith, P. & T. Nickles (eds) 1983. *PSA 1982. Proceedings of the 1982 Biennial Meeting of the Philosophy of Science Association*. East Lansing: Philosophy of Science Association.

Baker, G. & P. Hacker 1984. *Scepticism, Rules, and Language*. Oxford: Blackwell.

Barnes, B. 1982. *T. S. Kuhn and Social Science*. London: Macmillan.

Barnes, B. & D. Bloor 1982. Relativism, Rationalism, and the Sociology of Knowledge. See Hollis & Lukes (1982), 21–47.

Bird, A. 1998. *Philosophy of Science*. London: UCL Press.

Bird, A. 1999. Scientific Revolutions and Inference to the Best Explanation. *Danish Yearbook of Philosophy* **34**, 25–42.

Boyd, R. 1984. The Current Status of Scientific Realism. See Leplin 1984, 41–82.

Bloor, D. 1976. *Knowledge and Social Imagery*. London: Routledge & Kegan Paul.

Bloor, D. 1997. The Conservative Constructivist. *History of the Human Sciences* **10**, 123–5.

Bloor, D. 2000a. Wittgenstein as a Conservative Thinker. See Kusch (2000), 1–14.

Bloor, D. 2000b. Why Did Britain Fight the First World War with the Wrong Theory of the Aerofoil? Inaugural Lecture, University of Edinburgh, 8 May 2000.

Broad, W. & N. Wade 1985. *Betrayers of the Truth*. Oxford: Oxford University Press.

Brock, W. 1992. *The Fontana History of Chemistry*. London: Fontana.

Bruner, J. & L. Postman 1949. On the Perception of Incongruity: A Paradigm. *Journal of Personality* **18**, 206–23.

Carnap, R. 1928. *Der logische Aufbau der Welt*. Berlin: Benary.

Carnap, R. 1939. *Foundations of Logic and Mathematics – International Encyclopedia of Unified Science* vol. I, no. 3. Chicago: University of Chicago Press.

Carnap, R. 1963. Intellectual Autobiography. In *The Philosophy of Rudolph Carnap*, P. Schilpp (ed.), 3–84. La Salle, Ill.: Open Court.

Churchland, P. 1979. *Scientific Realism and the Plasticity of Mind*. Cambridge: Cambridge University Press.

Cohen, I. B. 1985. *Revolution in Science*. Cambridge MA: Harvard University Press.

Collins, H. & T. Pinch 1993. *The Golem: What Everyone Should Know About Science*. Cambridge: Cambridge University Press.

Colodny, R. (ed.) 1966. *Mind and Cosmos: Essays in Contemporary Science and Philosophy*. Pittsburgh: University of Pittsburgh Press.

Crombie, A. (ed.) 1963. *Scientific Change*. London: Heinemann.

Devitt, M. 1979. Against Incommensurability. *Australasian Journal of Philosophy* **57**, 29–50.

Doppelt, G. 1978. Kuhn's Epistemological Relativism: An Interpretation and Defense. *Inquiry* **21**, 33–86 [reprinted in Meiland & Krausz (1982), 113–46].

Doppelt, G. 1980. A Reply to Siegel on Kuhnian Relativism. *Inquiry* **23**, 117–23.

Earman, J. (ed.) 1992. *Inference, Explanation and other Frustrations. Essays in the Philosophy of Science*. Berkeley, CA: University of California Press.

Enç, B. 1976. Reference of Theoretical Terms. *Nous* **10**, 261–82.

English, J. 1978. Partial Interpretation and Meaning Change. *Journal of Philosophy* **75**, 57–76.

Feyerabend, P. 1962. Explanation, Reduction, and Empiricism. In *Scientific Explanation, Space, and Time: Minnesota Studies in the Philosophy of Science* vol. 3, H. Feigl & G. Maxwell (eds), 28–97. Minneapolis: University of Minnesota Press.

Feyerabend, P. 1975. *Against Method. Outline of an Anarchistic Theory of Knowledge*. London: New Left Books.

Fine, A., M. Forbes, L. Wessels (eds) 1991. *PSA 1990. Proceedings of the 1990 Biennial Meeting of the Philosophy of Science Association vol. 2*. East Lansing: Philosophy of Science Association.

Fischer, F. 1936. Review of Ludwik Fleck's *Entstehung und Entwicklung einer wissenschaftlichen Tatsache*. *Nervenarzt* **9**, 137–9.

Fleck, L. 1935. *Entstehung und Entwicklung einer wissenschaftlichen Tatsache: Einführung in die Lehre vam Denksdtil und Denkkollektiv*. Basel: Benno Schwabe. [For translation see Fleck 1979.]

Fleck, L. 1979. *Genesis and Development of a Scientific Fact*, T. J. Trenn & R. K. Merton (eds), F. Bradley & T. J. Trenn (trans.), foreword by T. S. Kuhn. Chicago: University of Chicago Press. [Translation of Fleck 1935.]

Frazer, J. 1890. *The Golden Bough: A Study in Comparative Religion*. London: Macmillan.

Fuller, S. 2000. *Thomas Kuhn*. Chicago: University of Chicago Press.

Grant, R. 1852. *History of Physical Astronomy*. London: R. Baldwin.

Gutting, G. (ed.) 1980. *Paradigms and Revolutions*. Notre Dame: University of Notre Dame Press.

Hacker, P. 1987. *Appearance and Reality*. Oxford: Blackwell.

Hacking, I. (ed.) 1981. *Scientific Revolutions*. Oxford: Oxford University Press.

Hacking, I. 1993. Working in a New World: The Taxonomic Solution. See Horwich (1993), 275–310.

Hanson, N. R. 1965. *Patterns of Discovery*. Cambridge: Cambridge University Press.

Hempel, C. 1935. On the Logical Positivist Theory of Truth. *Analysis* 2, 49–59.

Hempel, C. 1952. *Fundamentals of Concept Formation in Empirical Science*. Chicago: Chicago University Press.

Hempel, C. 1958. Theoretician's Dilemma. See Hempel (1965), 173–226.

Hempel, C. 1965. *Aspects of Scientific Explanation*. New York: Free Press.

Hempel, C. 1966. *Philosophy of Natural Science*. Englewood Cliffs, NJ: Prentice Hall.

Hesse, M. 1983. Comment on Kuhn's 'Commensurability, Comparability, Communicability'. See Asquith & Nickles (1983), 704–11.

Hiley, D., J. Bohman, R. Shusterman (eds) 1991. *The Interpretative Turn: Philosophy, Science, Culture*. Ithaca, NJ: Cornell University Press.

Hollis, M. & S. Lukes (eds) 1982. *Rationality and Relativism*. Cambridge MA: MIT Press.

Horwich, P. (ed.) 1993. *World Changes. Thomas Kuhn and the Nature of Science*. Cambridge MA: MIT Press.

Horwich, P. 1998 [1990]. *Truth*, 2nd edn. Oxford: Clarendon Press.

Hoyningen-Huene, P. 1990. Kuhn's Conception of Incommensurability. *Studies in History and Philosophy of Science* 21, 481–92.

Hoyningen-Huene, P. 1993. *Reconstructing Scientific Revolutions: Thomas S. Kuhn's Philosophy of Science*, A. T. Levine (transl.) with a Foreword by Thomas S. Kuhn. Chicago: University of Chicago Press.

Hoyningen-Huene, P., E. Oberheim, H. Andersen 1996. On Incommensurability. *Studies in History and Philosophy of Science* 27, 131–41.

James, W. 1904. Humanism and Truth. *Mind* 13, 457–75.

Kant, I. 1933 [1787]. *Critique of Pure Reason*, N. Kemp Smith (ed.). London: Macmillan.

Kitcher, P. 1978. Theories, Theorists and Theoretical Change. *Philosophical Review* 87, 519–47.

Koestler, A. 1968 [1959]. *The Sleepwalkers*. Harmondsworth: Penguin.

Kornblith, H. 1999. Distrusting Reason. *Midwest Studies in Philosophy* 23, 181–96.

Kripke, S. 1980. *Naming and Necessity*. Cambridge, MA: Harvard University Press.

Kroon, F. W. 1985. Theoretical Terms and the Causal View of Reference. *Australasian Journal of Philosophy* 63. 143–66.

Kroon, F. 1987. Causal Descriptivism. *Australasian Journal of Philosophy* 65, 1–17.

Krüger, L., L. Daston, M. Heidelberger (eds) 1987. *The Probabilistic Revolution*. Cambridge: Cambridge University Press.

Kusch, M. (ed.) 2000. *The Sociology of Philosophical Knowledge*. Dordrecht: Kluwer.

Lakatos, I. & A. Musgrave (eds) 1970. *Criticism and the Growth of Knowledge*. London: Cambridge University Press.

Laudan, L. 1981. A Confutation of Convergent Realism. *Philosophy of Science* 48, 19–49.

Laudan, L. 1984. *Science and Values. The Aims of Science and their Role in Scientific Debate*. Berkeley, CA: University of California Press.

Laudan, L. 1989. If it Ain't Broke, Don't Fix It. *British Journal for the Philosophy of Science* 40, 369–75.

Laudan, R. 1979. The Recent Revolution in Geology and Kuhn's Theory of Scientific Change. See Asquith & Hacking (1979), 227–39 [reprinted in Gutting (1980), 284–96].

Leplin, J. (ed.) 1984. *Scientific Realism*. Berkeley, CA: University of California Press.

Lewis, D. 1970. How to Define Theoretical Terms. *Journal of Philosophy* 67, 427–46.

Lewis, D. 1972. Psychophysical and Theoretical Identifications. *Australasian Journal of Philosophy* **50**, 249–58.
Lewis, D. 1983. New Work for a Theory of Universals. *Australasian Journal of Philosophy* **61**, 343–77.
Lewis, D. 1986. *On the Plurality of Worlds*. Oxford: Blackwell.
Lipton, P. 1991. *Inference to the Best Explanation*. London: Routledge.
Lyons, J. 1977. *Semantics*. London: Cambridge University Press.
Lyons, J. 1981. *Language and Linguistics: An Introduction*. Cambridge: Cambridge University Press.
Mannheim, K. 1953. *Essays on Sociology and Social Psychology*. London: Routledge and Kegan Paul.
Marx, K. 1913 [1859]. *A Contribution to the Critique of Political Economy*. Chicago: Kerr.
Meiland, J. & M. Krausz (eds) 1982. *Relativism Cognitive and Moral*. Notre Dame: University of Notre Dame Press.
Mellor, D. H. 1977. Natural Kinds. *British Journal for the Philosophy of Science* **28**, 299–312.
Miller, D. 1974. Popper's Qualitative Theory of Verisimilitude. *British Journal for the Philosophy of Science* **25**, 178–88.
Musgrave, A. 1971. Kuhn's Second Thoughts. *British Journal for the Philosophy of Science* **22**, 287–97.
Newton-Smith, W. 1981. *The Rationality of Science*. London: Routledge.
Niiniluoto, I. 1997. Reference Invariance and Truthlikeness. *Philosophy of Science* **64**. 546–54.
Nola, R. 1980. Fixing the Reference of Theoretical Terms. *Philosophy of Science* **47**, 505–31.
North, J. 1994. *The Fontana History of Astronomy and Cosmology*. London: Fontana.
Ortony, A. 1979. *Metaphor and Thought*. Cambridge: Cambridge University Press.
Pannekoek, A. 1961. *A History of Astronomy*. London: Allen & Unwin.
Papineau, D. 1979. *Theory and Meaning*. Oxford: Clarendon Press.
Papineau, D. 1993. *Philosophical Naturalism*. Oxford: Blackwell.
Papineau, D. 1996. Theory-Dependent Terms. *Philosophy of Science* **63**, 1–20.
Peirce, C. S. 1934. *Collected Papers of Charles Sanders Peirce*, C. Hartshorne & P. Weiss (eds). Cambridge MA: Harvard University Press.
Planck, M. 1949. *Scientific Autobiography and Other Papers*. New York: Philosophical Library.
Popper, K. 1959. *The Logic of Scientific Discovery*. London: Hutchinson.
Preston, J. 1997. *Feyerabend*. Cambridge: Polity.
Putnam, H. 1975a. *Mind, Language, and Reality: Philosophical Papers Vol. 2*. Cambridge: Cambridge University Press.
Putnam, H. 1975b. The Meaning of 'Meaning'. See Putnam (1975a), 215–71.
Putnam, H. 1981. *Reason, Truth and History*. Cambridge: Cambridge University Press.
Quine, W. 1953. Two Dogmas of Empiricism. In *From a Logical Point of View*, 20–46. Cambridge, MA: Harvard University Press.
Quine, W. 1960. *Word and Object*. Cambridge MA: MIT Press.
Quine, W. 1969. *Ontological Relativity and Other Essays*. New York: Columbia University Press.
Ramsey, F. 1990. *Philosophical Papers*, D. H. Mellor (ed.). Cambridge: Cambridge University Press.
Sankey, H. 1990. In Defence of Untranslatability. *Australasian Journal of Philosophy* **68**, 1–21.
Sankey, H. 1991a. Incommensurability and the Indeterminacy of Translation. *Australasian Journal of Philosophy* **69**, 219–23.
Sankey, H. 1991b. Translation Failure Between Theories. *Studies in History and Philosophy of Science* **22**, 223–36.
Sankey, H. 1993. Kuhn's Changing Concept of Incommensurability. *British Journal for the Philosophy of Science* **44**, 759–74.
Sankey, H. 1994. *The Incommensurability Thesis*. Aldershot: Avebury.
Sankey, H. 1997a. *Rationality, Relativism and Incommensurability*. Aldershot: Ashgate.
Sankey, H. 1997b. Incommensurability: The Current State of Play. *Theoria* **12**, 425–45.
Sankey, H. 1998. Taxonomic Incommensurability. *International Studies in the Philosophy of Science* **12**, 7–16.
Sardar, Z. 2000. *Thomas Kuhn and the Science Wars*. Cambridge: Icon Books.
Saunders, P. 1980. *An Introduction to Catastrophe Theory*. Cambridge: Cambridge University Press.

Savage, C. W. (ed.) 1990. *Scientific Theories*, Minnesota Studies in Philosophy of Science 14. Minneapolis: University of Minnesota Press.

Shapere, D. 1964. The Structure of Scientific Revolutions. *Philosophical Review* **73**, 383–94 [reprinted in Shapere (1984: 37–48)].

Shapere, D. 1966. Meaning and Scientific Change. See Colodny (1966), 41–85 [reprinted in Shapere (1984), c.5; reprinted in Hacking (1981), 28–59].

Shapere, D. 1971. The Paradigm Concept. *Science* **172**, 706–9 [reprinted in Shapere (1984)].

Shapere, D. 1984. *Reason and the Search for Knowledge. Investigations in the Philosophy of Science*. Dordrecht: Reidel.

Siegel, H. 1980. Objectivity, Rationality, Incommensurability and More; review of T. S. Kuhn: *The Essential Tension*. *British Journal for the Philosophy of Science* **31**, 359–84.

Stich, S. 1996. *The Deconstruction of the Mind*. New York: Oxford University Press.

Stove, D. 1982. *Popper and After: Four Modern Irrationalists*. Oxford: Pergamon Press.

Sulloway, F. 1996. *Born to Rebel: Birth Order, Family Dynamics, and Creative Lives*. New York: Pantheon Books.

Suppe, F. 1974. *The Structure of Scientific Theories*. Urbana, Ill.: University of Illinois Press.

Suppe, F. & P. Asquith (eds) 1976. *PSA 1976: Proceedings of the 1976 Biennial Meeting of the Philosophy of Science Association vol. 1*. East Lansing: Philosophy of Science Association.

Taylor, C. (ed.) 1959. *The Third (1959) University of Utah Research Conference on the Identification of Scientific Talent*. Salt Lake City, Utah: University of Utah Press.

Thomason, R. 1969. Species, Determinates, and Natural Kinds. *Nous* **3**, 95–101.

Toulmin, S. 1970. Does the Distinction between Normal and Revolutionary Science Hold Water? See Lakatos & Musgrave (1970), 39–47.

Van Fraassen, B. 1980. *The Scientific Image*. Oxford: Oxford University Press.

Watkins, J. 1970. Against 'Normal Science'. See Lakatos & Musgrave (1970), 25–37.

Wiggins, D. 1980. *Sameness and Substance*. Oxford: Blackwell.

Williamson, T. 1992. Inexact Knowledge. *Mind* **101**, 217–42.

Williamson, T. 1996. Cognitive Homelessness. *Journal of Philosophy* **93**. 554–73.

Williamson, T. 1997. Knowledge as Evidence. *Mind* **106**, 717–14.

Wittgenstein. L. 1922. *Tractatus Logico-Philosophicus*, C. K. Ogden & F. P. Ramsey (trans.). London: Routledge & Kegan Paul.

Wittgenstein, L. 1953. *Philosophical Investigations*. Oxford: Blackwell.

Worrall, J. 1988. The Value of a Fixed Methodology *British Journal for the Philosophy of Science* **39**, 263–75.

Worrall, J. 1989. Fix it and be Damned: A Reply to Laudan. *British Journal for the Philosophy of Science* **40**, 376–88.

Index

304

Thomas Kuhn

irrationality 26, 74ff., 89, 93f., 214, 216–18, 268, 270; *see also* symmetry postulate, Strong Programme

James, W. 232, 296
justification 246, 248, 252–8, 260, 261f.
J-J principle 246, 248, 252, 254, 258

Kant, I., Kantianism 124, 127–32, 141f., 232, 258, 270, 280, 290, 296
Kekulé, F. A. von S. 15, 61
Kepler, J. 33, 48, 99f., 102ff., 107, 114, 116, 284, 289
Koestler, A. 297
Koffka, K. 13
Köhler, W. 13
Knorr-Cetina, K. 270, 272
knowledge vii, 6f., 16f., 20, 14–20, 75, 104, 221f., 240, 246ff. 252, 258, 264, 254f., 258–60, 268, 270f., 272–5, 297; *see also* sociology of scientific knowledge
K-K principle 246, 248, 252, 258, 264, 297
Kripke, S. ix, 180, 182f., 279, 293

Lakatos, I. viii, 3, 263, 286
Lamb, H. 65
Langevin, P. 58
language, double language model 12–13
Latour, B. 270, 272
Laudan, L. 226, 263, 297
Laudan, R. 285
Lavoisier, A. 40, 57, 61, 98, 108, 110, 114f, 116ff, 122, 133, 179, 230
Levy-Bruhl, C. 17–20
Lewis, D. 186–7, 189, 206, 293
lexicon 68f., 166, 192, 194f., 218, 239, 241, 251, 286
Lipton, P. 297
Locke, J 8, 125f., 128, 251, 290
logic 4f., 90–95, 242, 262, 282
Löw, J. 19
Luther, M. 48
Lyons, J. 294

Mach, E. 9–13, 280, 282, 290, 298
Mannheim, K. 2, 19–20, 270, 276, 298
Marx, K. 19f., 270, 282
mass, Newtonian and Einsteinian conceptions of 164ff., 170–74, 178, 191, 207
Masterman, M,. 282, 283
Maxwell, J. C. 35, 41, 57, 66, 283
meaning 5, 11–13, 149–207
Mellor, D. H. 294, 298
Mendel, G. 285
Mendeleev, D. 57
mentality, primitive and civilized 17; *see also* Levy-Bruhl
Merton, R. 2
Michelson-Morley experiment 35, 283
Mill, J. S. 10

Miller, D. 292
More, L. T. 44
Musgrave, A. 76f., 287

natural selection 5f., 19, 211–14; *see also* Darwin
naturalism *see* epistemology
Needham, J. 270
neo-Platonism 43
network, lexical 192
Neurath, O. 232ff., 282, 287, 298
neutralism, epistemological *see* epistemology
neutralism about truth *see* truth
New Paradigm 3–9
Newton, I. 30, 33ff., 43, 45., 53, 57, 61, 71, 77f., 85ff., 122, 130, 156, 161, 164ff., 174, 189, 195, 225, 237f., 259, 284, 292, 296
Newton-Smith, W. 215f., 282, 291, 292, 295
no-overlap principle 197–202, 237
North, J. 283

observation 97–122, 145–6, 196, 239f., 278, 290, 291
observation-theory distinction 5, 145–6; *see also* theory-ladenness of observation
Old Rationalism, 3–9, 20, 65, 70, 72, 143, 149, 172, 210, 242, 162f., 170f., 272, 276, 281, 297

Pannekoek, A. 285
Papineau, D. 187, 189, 206, 293, 297, 298
paradigm viii, 23–7, 29f., 41–2, 47, 49, 50ff., 58–63, 65–96, 98, 110, 119, 122, 123f., 128, 130, 138–9, 140f., 143, 147–8, 149, 151, 165, 209, 241ff., 252f., 259f., 265, 269, 274, 279f., 282, 283
disciplinary matrix 68, 75–9, 81, 86, 244, 261, 282, 287
exemplar 68–71, 75–90, 93f., 106, 132, 140, 151, 165, 194ff., 213, 241ff., 245, 261, 287
as explanation 67–71
proto-paradigm 32
semantic function of 68f., 83–4, 119, 165
science without 30–2
shifts in 41, 50ff., 58, 76, 110, 122, 127f., 132–3, 147–8, 151, 153, 155f.
Pauling, L. 39
perception, perceptual experience 97–148, 196f., 215, 261, 290; *see also* seeing
gestalt 13–14, 19, 27, 84, 91, 100, 108, 111–19, 129
world change 123–33
phenomenalism 10f.
phlogiston, theory of 40, 44, 56, 108, 116ff., 184–5, 191, 230, 239
Pierce, C. S. 296